Matthias Thun (1948–2020) was an international expert on biodynamic beekeeping with over 50 years' experience. He had a Masters in Beekeeping and lectured at biodynamic conferences and training seminars internationally. He worked with his mother, biodynamic pioneer Maria Thun, on the annual creation of the seminal *Maria Thun Biodynamic Calendar*, until her death in 2012 after which he continued it independently. Matthias lived at his family's biodynamic farm in Dexbach, Germany, until his death.

Biodynamic
BEEKEEPING

A sustainable way to keep happy, healthy bees

MATTHIAS THUN

Floris
Books

Translated by David Heaf

First published in German as *Die Biene, Haltung und Pflege Unter Berücksichtigung Kosmischer Rhythmen* by Aussaattage-Verlag Thun & Thun OHG in 2015
Abridged version first published in English in 2020 by Floris Books, Edinburgh. Second printing 2022

British Library CIP data available
ISBN 978-178250-674-4
Printed and bound by MBM Print SCS Ltd, Glasgow

 Also available as an eBook

MIX
Paper from
responsible sources
FSC® C117931

 Floris Books supports sustainable forest management by printing this book on materials made from wood that comes from responsible sources and reclaimed material

Contents

Foreword *by David Heaf* 7

1. The Start of the Bee Year 11

2. Caring for Bees According to Cosmic Rhythms 19

3. The First Spring Inspection 31

4. The Building Frame 37

5. The Urge to Swarm 43
   The colony is allowed to swarm 46
   Preventing the swarm but retaining the young queens 49
   Preventing the swarm but swarm cells
     are not required for breeding queens 51
   Controlling and preventing the swarm urge 52
   The Marburg box 55

6. Colony Regeneration and Propagation 59
   Natural increase in colony numbers 61
   Prime swarm at the site of the parent colony 68
   Artificial colony increase 70
   Colony reproduction with bred queens 75
   Various options for colony reproduction 75

7. Breeding Queen Bees 88
   Breeding in queenless colonies 92
   Breeding in queen-right colonies 99
   Queen reproduction through deliberate use
     of the swarming instinct 101

8. Honeycomb Construction 103
    Natural comb construction 105
    Foundation 111
    The use of natural comb and comb
        built with foundation 111
    Changing over to natural comb 112
    Building in the honey chamber 114
    The age of foundation wax 115
    Conclusion 117

9. Honey 121
    Nectar 126
    Processing the honey 131
    Types of honey and their uses 138

10. Feeding in Winter 142

11. Bee Diseases 150
    Brood disease 152
    Adult bee diseases 158
    Diseases that harm both brood and adult bee 162

12. Methods of Ash Usage 169
    Potentising the ash 173
    The application of ground (dynamised) ash 175
    Varroa and drones 176

13. The Cultivation of Plants for Bees 180

14. The Conservation of Bees for the Future 185

Bibliography and Recommended Reading 189
Index 190

Foreword

It was with great sadness that we learned of the sudden passing of Matthias Thun in June 2020 while the English edition of this book was being prepared for publication. Matthias was the son of Maria Thun, the biodynamic pioneer and creator of the *Biodynamic Calendar*. He applied her methods to beekeeping and conducted his own extensive research and experimentation. The results of his invaluable work, carried out over the course of fifty years of caring for bees, are to be found in this book.

The translation of this book has been long awaited by English-speaking beekeepers who are interested in biodynamics. Not only is it the first book to translate into modern beekeeping practice Rudolf Steiner's indications on the essential nature of the honey bee, indications that have since inspired many apiarists to join the growing trend towards natural beekeeping, but it is also the only practical beekeeping book that integrates awareness of cosmic phenomena into bee husbandry.

Rudolf Steiner (1861–1925), the founder of anthroposophy, a science of the spirit, lectured on many aspects of practical life, one of which was an approach to plant and animal husbandry that is holistic in the fullest possible sense. It came to be called biodynamics, and today, biodynamic agriculture and horticulture take their place amongst the foremost initiatives engaged in organic food production, recognisable by the Demeter certification label.

Inspired by Steiner's indications in his agriculture lectures on the involvement in terrestrial life processes of not only the planets but also the constellations of the zodiac, Maria Thun (1922–2012) developed her *Biodynamic Calendar*, which Matthias subsequently applied to beekeeping. Depending on the position of the planets and the constellation, the days of the year acquire different qualities according to the four classical elements. Instead of disturbing the bees on days that suit the beekeeper's convenience, Matthias shows how, with the help of the calendar, the most propitious times can be chosen for carrying out such tasks as hive inspections, harvesting honey, artificial swarming and even grafting larvae.

A special feature of the author's method is working with the swarm urge rather than against it, as is common in modern beekeeping. He cites a vivid example of how long-term swarm suppression leads to colony disease. In fact, recent research has shown that when colonies are allowed to swarm naturally, the reinvigoration this brings contributes to their coping in the long term with the varroa mite and its highly damaging viruses.

Another feature is the use of natural combs, built by the bees themselves without the use of artificial foundation, which avoids introducing into the hives pesticides that have accumulated in old and recycled beeswax. We are given a thorough guide to the management of natural combs, which it is almost impossible to find in other modern beekeeping books. Matthias Thun describes the use of a special case of such combs, namely the building frame, a device almost unknown outside the European continent, which allows the beekeeper to judge the status of a colony without opening it. Opening a hive, thereby letting out the colony's warmth, must surely be among the most bee-unfriendly tasks that the beekeeper has to perform.

Following an indication by Steiner to add chamomile tea to bee feed to make it more digestible, Matthias extends this to include teas of all the other plants used in the biodynamic preparations: yarrow, chamomile, dandelion, valerian, nettle, oak bark and horsetail. This further helps link his approach to the biodynamic method, as it is impossible for biodynamic beekeepers to ensure that their bees' foraging range (often more than a mile in radius) is entirely on land under biodynamic management.

I commend this book to not only experienced beekeepers who would like to transition to a more bee-friendly approach, but also beginners who, while having access in some form to instruction about basic beekeeping, do not want to be steered into the conventional, mechanistic apiculture that new generations of beekeepers are increasingly finding unsatisfactory.

David Heaf
Author of *The Bee-friendly Beekeeper*

Chapter 1
The Start of the Bee Year

Introductory Note from the Publisher

Over the last hundred years, modern beekeeping has been transformed by new practices and new technologies, from the introduction of plastic building frames in hives to the artificial breeding of queen bees, all with the aim of increasing the production of honey and maximising profit. While these methods have been successful, they have proven detrimental to the health and well-being of the bees. Bee colonies are now weaker and more vulnerable to diseases such as varroosis, which has led to an alarming decline in the bee population and raised concerns about an impending ecological crisis. The future of the bee appears uncertain.

Faced with these worrying circumstances, beekeepers are beginning to ask if there is a more sustainable way of looking after bees, one that not only takes into account their natural habits and essential natures, but also considers their connection to their environment and the influence it has on them. This book explores one such method by describing the effects that cosmic rhythms – the movements of the planets and the stars – have on the activity of bees. Matthias Thun's methods are based on over fifty years' experience of beekeeping and on the numerous experiments and observations that he and the Thun family have carried out in this area. He has found, for example, that certain times are better for promoting the comb-building instinct and colony reproduction, while others are better for nectar foraging. Bees are more contented and easier to work with on certain days, whereas on others they are agitated and more liable to sting. This holistic method has also helped in combatting disease, allowing colonies to remain strong and healthy throughout the year.

While this approach may seem unconventional, for those beekeepers wanting to explore a more bee-centred, bee-friendly approach to caring for their bees it offers a harmonious way of working that is in accordance with the bees' essential nature, and one that seeks to preserve the bees for the future.

Matthias Thun begins his exploration of cosmic rhythms with a description of the bee year.

* * *

Most conventional beekeepers think of the bee year as beginning in late summer. According to their thinking, late summer care needs to be carried out conscientiously because only strong and healthy colonies will survive the winter.

This view of the bee year holds if we assume that winter bees only emerge from the last brood period of late summer, but scientists have discovered that winter bees also come from summer bees that have spared themselves from the industrious and exhausting labours of their fellow workers. This raises the question as to whether conventional opinion about summer or late summer care is tenable.

Other observations raise similar doubts. If we take a year in which there is extremely strong foraging in woodland during the first half of July (for example, when honeydew flow predominates), then late summer colony care might not yield the expected success, as I have observed from the relatively weak colonies that emerge the following spring. Winter may also present very varied weather conditions that can favourably or adversely affect the wintering bees. These examples of variation (and more could be added) suggest to me that in spring we cannot reckon with the colonies we wintered, but only with those that have survived the winter. We cannot assume that perfect management in late summer will result in perfect colonies in spring. This does not mean that such care should be neglected, but rather that we must make do and be satisfied with whatever colonies make it through the winter. The conventional emphasis on late summer care as starting the bee year needs to be challenged.

Looking again at research that shows that some summer bees are already earmarked for winter, it becomes clear that colony management over the whole of any year will be reflected in the following year. This means that some rethinking is necessary. We can no longer talk separately of summer bees and winter bees, because even in wintering colonies there are summer bees. The sense of separation between seasonal colonies and of the start of a brand new bee year in late summer leads us astray.

It is more appropriate therefore to think of the bee colony as an organism, *because that is exactly what it is.*

It is more appropriate therefore to think of the bee colony as an *organism*, because that is exactly what it is.

Austrian philosopher Rudolf Steiner, the founder of biodynamic agriculture, compared the organism of the bee colony with that of the human being. He indicated that the material part of the human physical body is replaced every seven years (*Bees*, pp. 67f). This turnover of substances does not happen instantly but occurs over the course of years. Looking at the organism of the bee colony, we can easily imagine that the turnover of summer and winter bees is likewise not sudden but is spread over the bee year.

Similarly, the transition of the winter colony to the summer colony is not an abrupt one but occurs over a period of time. The kink in colony development that you might sometimes observe in spring when the winter bees are suddenly absent tends to occur if the winter bees are poorly nourished, thereby curtailing their lifespan, and the development of the colony in spring is disturbed by unfavourable weather. In years when flower and honeydew forage are in a favourable ratio to one another, bees that are under-nourished rarely reach winter. This makes the kink in spring development too small to be noticeable. In normal weather conditions, and with healthy colonies, the transformation in a bee colony happens gradually and harmoniously.

In light of the foregoing, it is worth reconsidering the start of the bee year. If we use the conventional beekeeping concept of a 'bee year' that begins in late summer, we are actually ascribing to bees a different kind of time than that found in the rest of nature. The sun plays an incredibly important part in nature. In the spring, as the sun rises higher and the days begin to lengthen, nature prepares for new life in the summer; later in the year, as the darkness draws in, nature readies itself for winter. Why then should the bees be understood as an exception, especially as their life is strongly determined by this course of the year? In the lectures on bees that he gave to the workers at Switzerland's centre of anthroposophy, the Goetheanum, Rudolf Steiner explained that the queen and worker bees are sun beings. From this, it is clear that we should equate the course of the bee year with the course of the sun. The natural course of the year starts when the sun is increasing its daily arcs across the sky and ends when it has reached its lowest or deepest position. I will therefore begin my considerations of the bee year from midwinter after the solstice: the period of the sun's ascent.

At this start of the sun's year, beekeepers have very little work to do, besides checking the hive entrance to avoid it being blocked by too many dead bees. In hives with double floors, when the double floor has a second floorboard with an opening for climbing upwards, care needs to be taken that this opening is not blocked or else the bee colony will be disturbed due to lack of air. In the worst case it can cause suffocation.

I have observed that clearing flights may occur when the sun and Venus are in the constellation Aquarius* with additional Uranus storm aspects. So I would advise removing any anti-bird nets and mouse guards a few days before this, to ensure that the bees are not obstructed in their clearing flights. It is at just this time of year that the rays of the sun can bring about a very pleasant rise in the outside temperature, but if the bees enter a cool air current and try to fly home with their remaining energy, obstructions like nets can be fatal.

Since beekeepers worldwide have been faced with varroosis, I, like many others, have reintroduced the traditional winter inspection sheet (*Winterwindel*), albeit in modernised form. It is a cardboard insert on the hive floor on which the debris that drops during the winter can be examined. As with removing nets and obstructions, I'd recommend taking out this inspection

---

* The astronomical constellation of Aquarius is meant here, not the astrological sign. See *The Maria Thun Biodynamic Calendar* for details.

sheet one or two days before the clearing flights are expected, otherwise, at least with strong colonies, the debris will be carried away, thus denying you the opportunity to gain the information about the colony that it provides. The inspection sheet tells us about many things other than varroa. From the distribution of the debris, beekeepers can understand the position of the colony on the combs and its strength. The amount of discarded wax shows how much of the winter stores have already been consumed. The inspection sheet can thus tell the beekeeper many things that may otherwise be discovered only by opening the hive to view the colony.

By looking at the dead bees on the sheet before the clearing flights, the beekeeper can tell whether a colony is queenless (if there is a queen among the dead bees). If there are also drones among the dead bees, I would assume that the colony requeened itself very late on, and that the queen was no longer able to go on her mating flights, thus causing the colony to retain its drones for a long time. Such a colony would probably contain an unmated queen, evidenced by drone brood, or no queen at all because it became queenless in the course of the winter. All these examples show how useful a winter inspection sheet can be, beyond the check for varroa.

The day of the clearing flight is an exciting one for beekeepers. After many weeks of quiet in the apiary, an incredible commotion starts that is otherwise experienced only at swarming or peak foraging times. The colonies appear eager to emerge from all that intense tranquillity and summon up their strength for the new sun year.

My observations suggest that by looking more closely at the individual hive entrances, beekeepers can often find answers to questions arising from the winter inspection sheets. If, for example, a colony has a dead queen among the dead bees and the colony really is queenless, then the bees will crawl around on the front of the hive on the day of the clearing flight as if they are looking for something. This is a clear signal to the beekeeper, who can prepare to help the colony on the next warm day as described below.

If on the day of the clearing flight one or two colonies are not flying at all, we can listen to them in the evening. For those with good hearing, it suffices to put an ear close to the hive entrance to hear what is happening inside. For those who do not wish to bend down so far, a rubber tube can be used to listen instead. If you hear a deep humming, the colony is a latecomer waiting for the next warm day for its clearing flight. If it is quiet, I'd recommend knocking on the front of the hive with a finger to 'wake up' the colony. If there is still no humming and no clearing flight, it is likely that the colony hasn't survived the winter. To be sure, we open the hive to check. If the colony is dead, the hive should be cleared out immediately, or else at least close the entrance so that other colonies don't remove any stores still present in the hive. Listening to a colony in the evening can also tell you whether a colony is queenless if you are attentive to the tone and quality of the humming.

With the approach of spring and the steady rise in the outside temperature, we can reckon that in a short time the first pollen carriers will be bringing in pollens such as hazel, alder, poplar and willow. This usually occurs during the period when Venus is in Aquarius. The beekeeper should use this time to unite queenless colonies with nuclei or nucs (small colonies of bees created from larger colonies), or with young colonies. Uniting is very easy in this time just before spring, and there are hardly any losses. If the queenless colony is stronger than the nuc, then I would suggest removing the roof from the queenless colony, spraying its combs with thyme water from a spray bottle and laying a sheet of newspaper punctured with ten to fifteen holes (use a thin nail or ballpoint pen) over the occupied brood box. The holes in the newspaper should be placed above the centre of the colony. The nucleus is likewise sprayed with thyme water and placed on top. If the nuc box does not exactly fit the hive containing the queenless colony below, I'd suggest placing it in a honey super that fits the hive better. Then close the hive and wrap it up warmly. The bees in the upper and lower boxes make contact through the holes in the paper. They chew the holes wider and then the two colonies unite. A further check is only necessary if spring is so far advanced that an inspection is to be done on all colonies in the apiary.

FIGURE 1. *Shreds of paper deposited at the hive entrance suggests that the uniting of the two colonies was successful.*

Chapter 2
Caring for Bees According to Cosmic Rhythms

My family, the Thuns, became aware of the varied rhythms of the cosmos as a result of plant experiments and weather observations that we conducted from 1952 onwards. Our observations raised the question: is the life of bees subject to cosmic rhythms that are similar to or the same as those that we find in plant life and weather phenomena, or does the life of bees have its own cosmic rhythms? We put bees on the balcony of our house in order to observe their activity more easily throughout the whole day.

This soon showed that the bees' behaviour was not the same every day. There were days when bees returned with fat pollen baskets. When we did the nail test – gently pressing a thumb against the bee's abdomen – we found that the crop was empty. There were also days when the bees were in the full flow of honey foraging, but they were bringing in no pollen despite the warm, sunny weather. There were days when the bees did not fly much

FIGURE 2. In 1964, this Lüneburg mini-skep with a cast (also known as a secondary swarm, or after-swarm) helped us get the first results of our observations. It led to the building of our large experimental apiary.

and brought home neither pollen nor nectar, then suddenly there were days when the pollen and honey foraging lasted two or more weeks. In between, there were times when the otherwise peaceful bees darted like lightning through the air and if someone got in their way, they were sure to be stung.

After we had observed these various facts for two years, a picture emerged that allowed us to attribute certain behaviours to cosmic influences. To give just a couple of examples here: long periods of good pollen

FIGURE 3. Our first large experimental apiary was a rebuilt, rear-access frame-hive bee house equipped with trough hives.

foraging tended to occur when Venus was in the constellations of Aquarius, Gemini and Libra, whereas long periods of good honey foraging often took place when Mercury was in Aries and Leo. This is helpful because if beekeepers know when the first pollen foraging is likely to occur, they can time stimulatory feeding so that the first brood is sufficiently developed to use the pollen appropriately.

There were days that fell repeatedly outside the above-mentioned broader rhythms that still brought good pollen or honey foraging, but here it was possible to find a connection with the sidereal rhythm of the moon. With the passage of the moon through particular regions of the zodiac, the bees sometimes brought in nectar, sometimes pollen, and sometimes neither one nor the other. This reminded us of Rudolf Steiner's lectures on bees and a question put to him by a beekeeper:

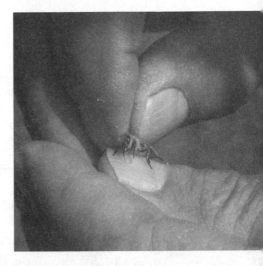

FIGURE 4. For the nail test a returning bee is caught at the hive entrance and her abdomen gently pressed with the thumb. If the bee contains honey or water in her crop, she vomits it onto the thumbnail.

> With regard to the influence that the signs of the zodiac can have on the production of honey, in some older, traditional farming communities, the farmers still pay close attention to this, for instance, when they sow the seed they do so when the moon is in Gemini and so on. Are they superficially following the characteristics of the signs of the zodiac, or is there something else that might be meaningful?

Steiner answered:

> Things of this type are naturally never properly examined in a scientific manner these days, even though it is possible to do so. I've already indicated to you what has an influence on the beehive. A bee, and specifically the queen, is to a certain degree a 'sun' animal. As a result, the sun, as it moves through the zodiac, has the greatest influence upon bees. The changes the sun undergoes as it moves through each sign are transmitted to the bee. But the bees are, naturally, also dependent on that which they find in plants. And this is where there is a connection between the farmers' sowing of the seed under a certain sign of the zodiac and the bees' finding of substances the plants have prepared for them. Such things aren't simply pulled out of nowhere; however, the way they are normally presented is very amateurish and misleading. These matters must be properly examined in a scientific manner so that they will have a firm foundation. (*Bees*, pp. 81f).

According to this it is a matter of what the bees find as substance in the plants. Does this mean that plants release nectar or pollen only under particular influences determined by their surroundings?

Lots of new questions suddenly arose. For example, it appeared that bee colonies also reacted very differently to the beekeeper working on them on different days. On some days the beekeeper can work undisturbed whereas on others their activity brings about real chaos amongst the colonies, which degenerates into a lively urge to sting.

With swarm-free methods, where colonies are prevented from swarming, colonies are inspected every nine days. This is determined by the developmental stages of the queen who goes from a laid egg to a capped cell in the same period. Depending on which day in spring the first inspection takes place, we then follow a pattern of activity that is dependent on sidereal lunar rhythms. These rhythms have been shown to influence plant growth and weather processes, and demonstrate a fourfold character that relates to the root, leaf, flower and fruit of the plant, and to the classical elements of earth, water, light (air) and heat (fire).

In specific inspection experiments we found that the bees' activity occurred in connection with the movements of the moon. If the moon passed through the constellations of Aries, Leo and Sagittarius, the bees brought more nectar, whereas if it passed through the regions of Gemini, Libra and Aquarius they engaged in pollen foraging. We also found that the bees' building impulse was stimulated when the moon passed through the regions of Taurus, Virgo and Capricorn. The bees live with their surroundings as if with a cosmic clock. Plants and the weather also resonate with this principle. For this reason, a wonderful harmony arises between plant and bee.

The bees live with their surroundings as if with a cosmic clock. Plants and the weather also resonate with this principle. For this reason, a wonderful harmony arises between plant and bee.

Constellation	Sign	Element	Microclimate	Plant	Bee
Pisces	♓	Water	Watery	Leaf	honey processing
Aries	♈	Heat	Warm	Fruit	nectar foraging
Taurus	♉	Earth	Cool/cold	Root	comb building
Gemini	♊	Light	Airy/light	Flower	pollen foraging
Cancer	♋	Water	Watery	Leaf	honey processing
Leo	♌	Heat	Warm	Fruit	nectar foraging
Virgo	♍	Earth	Cool/cold	Root	comb building
Libra	♎	Light	Airy/light	Flower	pollen foraging
Scorpio	♏	Water	Watery	Leaf	honey processing
Sagittarius	♐	Heat	Warm	Fruit	nectar foraging
Capricorn	♑	Earth	Cool/cold	Root	comb building
Aquarius	♒	Light	Airy/light	Flower	pollen foraging

TABLE 1: *The constellations of the zodiac, their correspondence with the four elements and the various parts of the plant, and the effect they have on the activities of bees.*

By working with bees on specific days, we not only influence and guide their activities, we can also help them strengthen their constitution according to the effects of certain constellations. Some recommendations for practical work according to this system are given opposite.

Light/flower times

The impulse given by light/flower times, when the moon is in the constellations Gemini, Libra and Aquarius, supports all work that is intended to promote colony build-up, colony reproduction and brood activity. Pollen foraging also increases at these times. The colonies remain very calm when worked on. They stay quietly on the combs, covering the brood without any agitated running around. Honey yield is above average.

Water/leaf times

At water/leaf times, when the moon is in the constellations Pisces, Cancer and Scorpio, it can frequently be observed that bees do not want to be disturbed. Their displeasure at being worked with on these days is evidenced by an increased tendency to sting. They are very agitated and move down off the combs as soon as the combs are removed from the hive. The honey yield is below average.

Warmth/fruit times

Colonies worked on during warmth/fruit times, when the moon is in the constellations Aries, Leo or Sagittarius, tend to exhibit increased nectar foraging. Through this one-sided compulsion, colonies produce very good honey yields in the first half of the year, but then largely neglect their work on the brood with the result that in the second half of the year they are to some extent no longer capable of peak yields. The neglect of brood is caused by these colonies bringing in less pollen than at light/flower times, thus creating a reduced food supply. During colony inspections they appear very calm and contented.

Earth/root times

Working colonies at earth/root times, when the moon passes through the constellations Taurus, Virgo and Capricorn, promotes the comb-building instinct. This shows particularly clearly when, for example, artificial swarms made at warmth/fruit times are left to build at earth/root times. Then, under the earth/root impulse, they accept a new organism which especially promotes the comb-building instinct. The honey yield is below average at these times. The colonies are not so gentle as those worked with during light or warmth times.

In order to make full use of the beneficial effects of these four impulses, we should proceed as follows:

Spring inspection of colonies is carried out at light/flower times. If a honey flow begins, we change to warmth/fruit times as these promote nectar gathering. If a pause in foraging occurs, we return to working colonies at light/flower times in order to stimulate their enthusiasm for raising brood. Light/flower times encourage brood activity. At the next nectar flow we change to warmth/fruit times.

We take advantage of earth/root times when we want to encourage the instinct to build. When we need lots of new combs, we give foundation* or building frames at earth/root times. The stimulus to colonies to build is then increased. The following light/flower time is omitted and we wait until the next light/flower time to work on the colonies again. This returns the bees to a more favourable rhythm and they can fully enjoy their comb-building drive.

The beekeeper should use water/leaf times for preparations and leave the bees completely alone. The bees are content with this and the beekeeper avoids unnecessary stings.

As well as the moon there are the planets that also shape the life of bee colonies. These stand in a particular relationship to the constellations and to the classical elements.

---

* If a comb is cut, we can see cells facing in opposite directions. These cells are separated in the middle by a septum or wall that comprises the floors of the cells. These cell floors are joined together in what is called the foundation.

Constellation	Element	Planet
Aries, Leo, Sagittarius	Warmth	Saturn, Mercury, Pluto
Taurus, Virgo, Capricorn	Earth	Sun, Earth
Gemini, Libra, Aquarius	Air/light	Venus, Jupiter, Uranus
Cancer, Scorpio, Pisces	Water	Mars, Moon, Neptune

TABLE 2: *The signs of the zodiac and the planets grouped according to their correspondence to the classical elements.*

Increased pollen foraging that enables good colony growth can be observed when Jupiter passes into light constellations. Through these observations we can predict when pollen and nectar foraging are due to begin. This can help facilitate our preparations for the bee season, from the breeding of queens to the harvesting and processing of the honey.

Beekeepers who breed queens know that some grafts behave very differently from others. In one case thirty-five queens resulted from one grafting session, but in another the same number of larvae produced only ten. To find out whether queen breeding is also dependent on the constellations, breeding comparisons were carried out over several years. It turned out that the most favourable time for the larval grafting to occur is when the moon and Venus are in a light constellation.

FIGURE 5. *The zodiac divided into trines with each sign allocated to one of the four elements.*

The honey harvest and its further processing should also be done under the influence of certain constellations. Warmth/fruit and light/flower times are best for this. Honey harvested at these times shows the finest setting and has the best taste. The decisive intervention is the removal of the combs from the colony.

The question as to how the bees receive the cosmic effects cannot as yet be answered with ultimate certainty. We know only that the beehive has to be opened in order to be able to activate these influences. From plant cultivation we know that cosmic forces can only take hold when the soil is moved. This movement has a chaotic effect on the organisms in the soil, and it is during the ensuing reorganisation that the effects of cosmic forces are taken up. A similar thing also happens in bee colonies when the hives are opened. By opening the hive, the protective layer of propolis is broken and light is let into the hive, thus causing a disturbance. In this way it can be assumed that cosmic effects gain access to the colony.

Figures 6 and 7. Cosmic forces become effective when opening a colony or hoeing the ground.

The favourable times for the beekeeper's work can be found in *The Maria Thun Biodynamic Calendar.* This annual publication gives favourable times not only for plant cultivation but also for beekeeping.

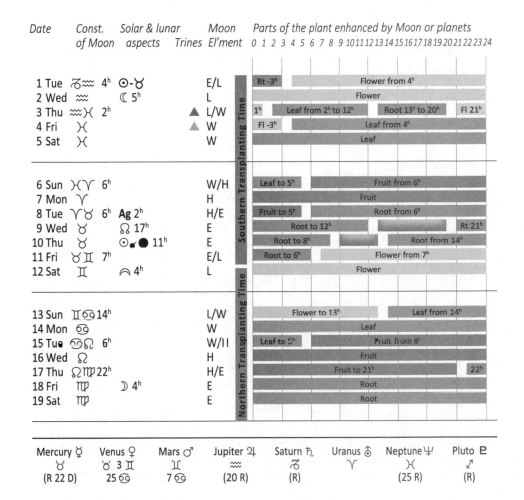

Date	Const. of Moon	Solar & lunar aspects	Trines	Moon El'ment	Parts of the plant enhanced by Moon or planets

FIGURE 8. The Maria Thun Biodynamic Calendar. *The calendar shows (on the left after the date) the constellation of the zodiac in which the moon is, and the time when the moon moves into each new constellation. The single letters show the element that the moon influences: Earth, Water, Light (or air) and warmth (H for Heat). The bars show the part of the plant enhanced, and the times when this effect lasts. The bars left blank indicate unfavourable times, usually caused by the nodes of the moon or planets (that is, when their paths cross the sun's path, the ecliptic). At the foot of the page the position of the planets in the zodiac are shown.*

These comments are intended to show how the beekeeper applies and uses *The Maria Thun Biodynamic Calendar*. Further explanations are included in the calendar.

Chapter 3
The First Spring Inspection

The first inspection of colonies in spring should be done at light/flower times (see previous chapter and *The Maria Thun Biodynamic Calendar*) when pollen is naturally available, as these support colony development. Some beekeepers, however, cannot wait, and as soon as spring arrives they inspect their colonies when there is still no pollen around. Going to the colonies so early is of little use: the bees are disturbed unnecessarily and they use up a lot of energy restoring the thermal balance in the hives. The right time for first inspection is when willows are in flower. The temperature is then generally such that colonies will suffer no harm through inspections.

Attention should be given to whether the colonies have brood at all stages, that is eggs, larvae and capped brood. If there is not yet any capped brood, this should be noted on the hive record card. It can be seen from the capped brood whether the queen is producing worker or drone brood. If the capped brood consists solely of clearly raised cells in which drones are developing, it can be assumed that she is an unmated, supersedure queen from the previous autumn. She must be removed from the colony and the colony rejuvenated with a nucleus or a young colony from the previous year. Apart from the brood, the broodless comb surfaces should also be inspected. As queens lay very vigorously in the spring, they need unrestricted access to sufficient empty comb so that the colonies can expand quickly. If there is insufficient empty comb it is advisable to hang a light-brown, empty comb, moistened with honey, at the edge of the brood nest. A colony expanded in this way, however, should occupy as far as possible all the combs of their winter cluster.

If we are dealing with a magazine hive – a type of hive that cannot be expanded with single combs – then so long as sufficient stores are present, a full comb may be replaced with an empty one.* If the colonies have consumed very little of their honey over the winter, we can encourage them to do so by scratching the capped surface with an uncapping fork. This will induce the bees to move it so as to make more space in the brood nest, as well as stimulate laying. However, not more than two or three hand-sized capped surfaces should be scratched as the honey in scratched cells can leak. If colonies do not quickly deal with the honey, it can trigger a tremendous gluttony in the apiary that can easily turn into robbing.

If colonies are suffering from a shortage of honey, they can be given stockpiled honeycombs or fed in the evening with 5 litres (about 5 quarts) of syrup made from honey or sugar in the proportions 1:1. For example, 1 kg (2 lb) of sugar to 1 litre of

* This book covers two basic hive types. One is magazine hives, those with stacked boxes such as Langstroths and UK Nationals, and the other is trough hives, which are more traditional in Germany and Eastern Europe. Magazine hives are extended vertically by adding more boxes, with each box possessing a more or less fixed number of combs. Trough hives are extended horizontally by adding more combs to the trough. Although trough hives are relatively rare in the UK and the USA, where they are generally referred to as 'horizontal hives', there is a growing interest in them, mainly among hobby beekeepers.

water (1 quart). In spring, honey stores should comprise at least 2½–5 kg (5½–11 lb) per colony, because there can still be breaks in the weather during which the bees may not be able to fly for weeks. Beekeepers for whom the main flow is early in the year or in early summer, may now stimulate their colonies with a hand-sized lump of fondant (a sugar-honey mixture). Feeding in such a way leads colonies to believe they have a forage source. This in turn increases brood production, and they are less likely to suffer the kind of setbacks that could lead to a shrinking of the brood nest as a result of bad weather.

After nine days, at the next light/flower time, the stronger colonies may be given a building frame. This is a small, empty frame with a strip of foundation 2–3 cms (1 in) wide affixed with molten beeswax on the top bar. The building frame is put in the hive to provide the colony with the opportunity to build drone comb. It also provides a very neat and clear way of reading the condition of the colony without having to inspect all of the brood combs every time; such inspections disturb colonies too much and increase the risk of unintentionally injuring the queen when moving the combs. The building frame should be as big as a brood comb. This is because the traditional small building frames, which in rear access hives are for viewing through a window, are built up too quickly; we can only read from them the colony status that prevails during the build-up phase. The most favourable option is when there are two drone comb frames in the colony and comb is cut from one of them each time it is inspected. This extends the build-up time and thereby gives the comb-building in those frames a more diagnostic value.

Since the arrival of the varroa mite, beekeepers have often been advised to cut out drone brood. This is certainly beneficial, but unfortunately some beekeepers now think they have to get rid of *all* drone brood from the colony in order to reduce varroa mites.

However, if we take such drastic action, we will soon notice that the bees are not as easily controlled as before, and that the incidence of disease also rises. I believe we should therefore leave a small amount of drone comb in our colonies. The drones belong to the organism of the bee colony just as much as the queen and workers.

After another nine days, when the next inspection of the colonies takes place, the building frame shows whether the colony is undergoing harmonious expansion or whether it is already in a swarm mood. In the case of a strong colony, the building frame will be fully built-up and there will be brood in it – eggs and larvae. To such a colony we then give a second building frame. In any apiary there are colonies of different strengths. We can therefore remove one or two freshly capped brood combs with bees from the strong colonies and use them for strengthening others. These brood combs are carefully inspected to ensure that the queen is not also being removed from the colony. Freshly capped brood is recognisable by it still having open cells with fat larvae, while in the middle of the comb all the cells are capped. The combs are placed on a comb rack and covered with a dry linen cloth or a close-weave cotton cloth to keep them warm.

After another nine days, when the next inspection of the colonies takes place, the building frame shows whether the colony is undergoing harmonious expansion or whether it is already in a swarm mood.

The colony from which the brood combs have been removed are given either two light-brown combs moistened with honey, or a light-brown, honey-moistened comb and a frame of foundation. These are inserted at the edge of the brood nest. This enables us to obtain nearly equal strengths among the colonies relatively early on.

By now the surrounding vegetation has increased, some orchard trees are in blossom and dandelions have started to flower. Because nectar flows at this time may already bring small honey yields, we switch our inspections from light/flower times to warmth/fruit times. The impulse of working colonies at warmth/fruit times causes colonies to concentrate more strongly on nectar foraging. At the coming inspections of the colonies, we increasingly observe and steer any swarming that may arise. Strong colonies that can tolerate more space are given it. This additional space could be used for brood expansion or for honey storage. Beekeepers with an extremely early flow make this space inaccessible to the queen by using a queen excluder, because they have to harvest the honey quickly.

Dandelion honey should not remain too long in the hive or it may begin to granulate in the cells. This danger arises if there is no queen excluder and the queen extends her brood nest into the extra space given. As brood combs should not be centrifuged, we have to wait until the brood has hatched. In the interim, the honey can thicken so much that it is no longer possible to spin it out of the comb. I would advise those who find the early flow plays a minor part to omit the queen excluder at this time to allow the colony to develop freely. Those who do not want to omit the queen excluder should, at the given time, move brood from the brood chamber into the honey chamber so as not to hinder further growth of the colony. In moving brood, care must always be taken not to move the queen into the honey chamber.

It is questionable whether it is right to remove brood from a growing brood nest and put it in a totally different part of the hive. The brood nest is begun and then extended or reduced as the colony desires it. The beekeeper might be justified according to their own ideas when they move combs, but certainly not according to the habits of the bees if this brings about a disorder in colony growth. If the bees do not want to swarm, they will probably prefer to extend the brood nest harmoniously and to do so without the hindrance of a queen excluder.

Chapter 4
The Building Frame

We are now in spring. The first forage plants, such as willow, are in flower. Colonies are becoming visibly larger and the time has arrived to put in the building frame. As has already been indicated, the building frame can tell the beekeeper a great deal and is very useful in colony management. At the time of rear-access hives, windows were integrated into the building frame. The beekeeper opened the hive door, or flap, and removed the insulating quilt to gain a clear view of the building frame through the glass. This gave the beekeeper the possibility of observing activity in the colony without opening the hive. Often the queen could be observed laying eggs, which could be a very instructive experience for beginners.

As top-access hives, especially magazine hives, became increasingly fashionable, the building frame was forgotten. At about the same time, a strange idea spread amongst beekeepers. They said you could increase the magazine arbitrarily, which would mean that swarming would no longer happen. The building frame could therefore be dispensed with completely. This went so far that many beekeepers *excluded* drone comb from colonies and gave them worker comb instead. As a result, the colonies would be stronger and make better yields possible. However, these views led to a kind of dead end. For one thing, the beekeeper had to realise that bees that did well in their rear-access hives failed to bring the promised success in magazine hives. The bees did not respond so well to more space being made available and continued to become swarmy. To the amazement of beekeepers, the swarm urge was now significantly more difficult to control than was the case in the rear-access hives.

It was a long time before the reason for this problem was discovered. Beekeepers needed to learn that, in hives such as magazine hives with lots of space, they needed bees that accepted and filled the large space available without first getting into a swarm mood. On the other hand, it was realised that colonies with drones were significantly easier to control, and this led to the reintroduction of drone comb or building frames. Indeed, for several years the control of varroosis shows that the building frame and drone comb have become a must with almost all beekeepers.*

The building frame is an empty frame the same size as brood frames. As a starter strip, a strip of foundation 3 cm (1 in) wide is affixed to the top bar with melted beeswax, or we take a comb and cut it out so as to leave a row of 3–5 cells along the top bar. We have to give the bees the beginnings of comb so they do not build haphazardly. These cells are bevelled to the septum in order to give the bees a fresh start. But if a piece of embossed foundation is affixed, we must make sure that the cell-wall ridges stand on their tips and not on their sides, as otherwise this would introduce a disharmonious element into the hive: the comb cells would not be constructed correctly and this would lead to a false assessment of the colony later on (see Figure 9 on page 40).

* The use of the building frame is rare or absent among anglophone beekeepers.

The building frame is hung at the edge of the brood nest. I recommend that they are put in the same place in all colonies, or marked with a colour or a marker pen, so that when inspecting the colony they can be seen immediately each time. This avoids disturbing the colony unnecessarily. I always situate the building frame as the penultimate frame at the back or side. This raises the possibility that, if small honey flows occur, the honey goes into the comb beyond the building frame, thus enabling a small harvest to be taken early on.

Colony inspections are done at light/flower times in the spring. If we insert the building frame then, and the weather promotes colony growth, then in nine days' time at the next inspection, the building frame should be at least half built. With very fast colonies it can already be filled, but it can sometimes happen that bees are hanging in the frame even though no building has taken place. If other colonies in the apiary are already building in their frames, this would indicate that this particular colony is lagging behind. There could be several reasons for this. The colony maybe weaker than the others. If so, we can strengthen it with combs of hatching brood and bees from strong colonies. Or the colony may have an old queen who is no longer able to support a vigorously growing colony. If the queen looks obviously worn out as a result of being constantly nibbled at by the bees, we can assume that there is no harmony in the colony because the queen is no longer fully accepted by her bees. In cases like this, it is a good idea to rejuvenate such colonies with young ones or nuclei (nucs) from the previous year.

Now we examine the built-up building frames a bit more closely. Cell construction should be absolutely perfect. This means that the cells should all stand on their tips (cell corners) and not be tilted in some way as has already been mentioned above.

> *There should be no mixture of different types of cells – no interweaving of worker and drone cells – as this is caused by a lack of harmony in the colony.*

Furthermore, the shape of the comb's perimeter should be harmonious. Observing this is possible only when the combs are not yet fully built. There should be no mixture of different types of cells – no interweaving of worker and drone cells – as this is caused by a lack of harmony in the colony. This lack of harmony may arise either from the relationship of the workers to the queen, or from serious mistakes by the beekeeper – for example, if you have accidentally inserted one or two frames of foundation in the brood nest and thus divided the intact brood nest into two parts.

FIGURE 9. *Left: Correct cell construction. Centre and right: Incorrect cell construction.*

While working through the hives, the building frame combs should all be cut out, regardless of how far they have been built up in individual colonies. After a further nine days, the building frames are inspected again and, to colonies that have already built up half of the frame area, a second building frame is given. The first building frame will now become drone comb that will always stay in the colony so the colony can regulate its own requirement for drones. The second building frame now takes the place of the first one. The first building frame, now permanently devoted to drone comb, should be placed at the side and spaced somewhat further from the neighbouring combs. This is because drones, being significantly longer than workers, need deeper cells in which to develop and thus require more space on either side. If this is not observed then it can happen that in spring, just when the colonies attach great value to drones, the worker combs adjacent to the drone combs are chewed back, thus taking away large areas of comb that could be laid in.

After a further nine days the building frame is again inspected. If in the interim swarm time has begun and some colonies are already in swarm mood, we can easily see this on the building frame. We remove the building frame and brush the bees off it so that a close inspection can be made. Our first focus is on the question of how far down the bees have built. If the comb is fully constructed then we concentrate on the brood, as shown in the previous chapter. In the centre of the comb we already see larvae; in the remaining part of the comb the cells will contain eggs.

If the larvae lie dry in the bottoms of their cells, this colony is not yet in swarm mood. If the larvae are deep in brood food, we should turn our attention to the brood nest, and carry out random checks to see whether there are already queen cups with eggs in them.

If the building frame is not already filled and the perimeter of the built comb is irregular, with perhaps an extension in one place in which a queen cup has already been placed, then we can assume that this colony is in a swarm mood. Now the beekeeper has to decide whether they would like to have the swarm issue or prevent swarming. Furthermore, they must decide whether they want to get young queens from this colony or whether they will carry out queen breeding.

Chapter 5
The Urge to Swarm

Various signs tell the beekeeper when the swarm mood is building in a bee colony. When colonies are inclined to swarm, they start building queen cells. Admittedly these are only the beginnings of cells, the so-called queen cups. These are like hemispherical cups built at the edges of combs all around the brood nest, and they indicate the beginning of the period in which the swarm mood may occur. This does not necessarily have to be in the spring; it can also extend into summer. The main swarm time is influenced by particular constellations and is thus not restricted to a particular time of year. These constellations occur when Venus or Mars is in Taurus. As these times change from year to year, there are years in which swarming is almost absent and limited to only a few, very swarm-prone, colonies.

Another variant is *weather-dependent swarming*. I mention the weather here because weather phenomena can likewise be put down to cosmic rhythms. It occurs in spring when foraging has got off to a good start; the colonies are bringing in lots of pollen and nectar and are growing rapidly. If such a foraging period is then interrupted by a prolonged bout of bad weather, colonies easily get into a swarm mood, even if the main swarm time has not yet started. When this happens, it is easy to get the impression that, because the colonies suddenly do not have much to do, they transfer their boredom to swarming.

A further trigger is *forage-dependent swarming*. This means that the swarming impulse can be caused by a period of very strong honey or pollen foraging. Included among the swarm-promoting forage plants are the crucifers such as rape and mustard, but also buckwheat and dandelion to some extent.

We can therefore include among the causes or triggers of swarming, both cosmic rhythms and intense foraging activity.

A good examination of the brood is necessary for detecting the beginning of the swarm mood. To this end, the *open brood* (all brood stages up to capping) is inspected and the quantity of brood food in which the larvae are floating is assessed. To train one's eye for these inspections, it is best to compare two colonies. One of them should evidently be in a swarm mood, as shown by the fact that there are already queen cells with eggs and larvae. The other should not be in a swarm mood. For the latter, we should choose a colony that has a young queen, as such a queen does not as a rule get into a swarm mood. If the young, small larvae in the pre-swarming hive are carefully examined, we notice that they are wholly immersed in milky brood food. Indeed, the entire cell floor is covered with it. Comparing this now with the colony that is not in swarm mood, we discover that its larvae are almost dry. They have significantly less brood food at their disposal.

If we want to check if a swarm mood is building in a colony, we need only assess the quantity of brood food in which the larvae float. If the larvae are still very dry, we no longer need to concern ourselves with swarming in that colony. On the other hand, if they are lying in plentiful brood food, we can assume with a high degree of certainty that the colony is in a swarm mood.

We can justifiably ask why colonies that have entered a swarm mood feed their larvae more generously than those colonies not in a swarm mood. To answer this, we need to look more closely at the emergence of the swarm mood. Across the various breeds of bees, we can differentiate between two groups: one that is very much inclined to swarm, and another that is extremely resistant to swarming. But the cause of the emergence of swarming is the same in both groups, and we find it in the ratio of the number of nurse bees to the number of larvae needing to be fed.

If a colony has significantly more nurse bees than larvae, then the larvae get fed more brood food. The nurse bees have to give away the brood food that is formed in their glands as they do not apparently have the ability to suddenly stop producing it. A similar phenomenon occurs in cows. Cows that are constantly milked have fully working milk glands and must be milked again within a certain period of time. If the milking is stopped, the cow continues to produce milk, but because it can no longer contain it, the cow lets the milk flow out. The cow may even get mastitis. A similar situation seems to apply to bees at the level of brood care, except that in this case the nurse bees do not simply let the brood food leak from their tongues, but instead overload the larvae. If all larvae have excess food, the bees apparently assume that the space occupied by the colony has become too small and so they steer the colony towards swarming.

To counter this, we can give our colonies additional space so that the queen can expand the brood nest. We can tell from the result of such expansion if we have swarming-resistant or 'swarmy' bees. We can give swarmy bees as much space as we like, but it will not bring them out of a swarming mood, whereas swarming-resistant bees can be influenced by this expansion of the brood nest – but only if the swarming mood is in its early stages, when only excessively fed worker larvae are present and the queen cells are still empty of eggs. If there are already one or two eggs or queen larvae present, then this colony will no longer respond to being given more space. The intention to swarm is then so strong that we must either let the colony do so, or undertake a major intervention in the colony's development.

The colony is allowed to swarm

All brood frames are inspected to see how many queen cups have eggs. As we cannot examine how old the eggs are, we must assume that they are three days old, the oldest they can be. If there is already a freshly hatched larva in a cell, we can assume that four days have elapsed since laying. What do these numbers tell us? For a queen it takes fifteen to sixteen days from egg laying to hatching (eclosion). This is divided as follows: three days for the egg stage, five to six days for the larval stage, and seven to eight days for the capped cell stage. The differences in duration may arise from large changes in temperature during this time. Swarm cells are placed, as a rule, in the marginal regions of the comb. If there is a sudden deterioration in the weather during which the outside temperature falls steeply, it can happen that some cells are not kept as warm and slight delays in development occur.

Returning to our swarming colony, we have found only eggs in the queen cells and have assumed that three days have elapsed since they were laid. The colony will be able to swarm from the tenth day onwards if the weather is correspondingly good, as the tenth day is the day when the queen cells are capped. In this situation, in order to take the swarm, the beekeeper will go to the apiary six days after the inspection in which eggs were found in the queen cups. If the colony does not swarm because of the weather, then the beekeeper has to return to the apiary on the seventh or eighth day. They do not have to stand guard from morning to night, because swarms generally issue between 10.00 am and 1.00 pm. Only when there are several days of bad weather will a swarm issue in the afternoon. With swarming it is similar to the mating flight of the queen – she takes care to select fine, sunny weather.

FIGURES 10–13. *Top left: Queen cup which clearly shows the sun-like round form. Top right: Part-extended queen cell. Bottom left: Hidden queen cell. Bottom right: Emergency queen cells on a worker comb.*

We can now ask why colonies swarm after the first swarm cell is capped. The answer usually given is that a colony would then be sure that the part of the colony staying behind will get a new queen. The old queen of course flies away with the swarm. Rudolf Steiner viewed this differently. He indicated that an egg laid in a queen cup begins to shine. This shining, although invisible

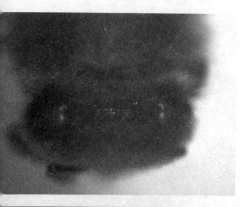

FIGURES 14 AND 15. Two dorsal ocelli (top) and one medial ocellus (bottom).

to humans, can nevertheless be seen by the bees. Apart from their compound eyes, bees have three small, simple eyes, known as ocelli, situated on their forehead above the bases of their antennae. Not much is known about these ocelli. Some say they are vestigial eyes. Experiments for many years have shown that covering the ocelli makes the bees fly out later in the morning and complete their flying earlier in the evening. From this it has been concluded that the ocelli are necessary for using the twilight phase in nature for flying.

Rudolf Steiner said that the bees can perceive the shining of the eggs with their ocelli. He indicated that this light is intolerable for the bees. From the egg then comes the larva that is capped on the ninth day. During this time, the shining of the developing queen decreases from day to day, becoming less intense until the cell is capped and the bees embark on the liberating experience of swarming (*Bees*, pp. 12–14). There are of course many queen cells in a colony intent on swarming. As a colony can issue not only one swarm but several, its seems a particular part of the colony adopts a particular swarm cell each time, and only then swarms together with the queen it has chosen.

Preventing the swarm but retaining the young queens

The colony that is in a swarm mood is checked again for the age of the queens developing in it, and the probable day of swarming is calculated. On the day before the swarm issues, the whole brood nest is removed from the colony. Light-brown brood combs are inserted into the empty brood chamber, and a capped brood comb with the queen on it is added to them. The bees that were on this brood comb should be left on it, but there must be no swarm cells or queen cups on it. To be on the safe side, the honey chamber should also be checked for queen cells and any that are present removed. The removed brood nest now has no queen and is made into nucs. The nuc with the queen should, however, remain in position because the flying bees returning to the old colony site can benefit the nuc; it may still produce a good honey yield even if the whole brood nest with the young bees has been removed. This method is one from a variety of swarm prevention methods.

There are beekeepers, however, for whom it is too inconvenient to wait a few days after finding queen cups with eggs in them before making a nuc from the brood nest.

A case has become known where a beekeeper chose the simplest route to queen renewal. He established that his colony was in swarm mood and had already found queen cups with eggs. Then, without further ado, he took the old queen from the colony, and could rest assured that this colony would no longer issue a swarm – an extremely easy way of getting rid of the swarm mood. But after several years, a nasty disease spread through this beekeeper's apiaries, namely *chalkbrood*. The colonies were so inwardly weakened that this disease reached epidemic proportions. Experts were called in to diagnose the problem, but they were unable to discover how this epidemic of chalkbrood could have arisen. However, after carefully talking through the management methods with this beekeeper, Ernst Perkiewicz (an experienced beekeeper and biodynamic practitioner) uncovered the reason. It lay in the fact that this beekeeper had, through premature interruption of the swarm urge, brought disharmony into the colonies, and by repeating this over many years had weakened the constitution of the colonies to the extent that the chalkbrood pathogen found the best conditions to grow.

How can we understand why such disharmony arises in bee colonies through prematurely interrupting the swarming instinct?

We may find an answer to this if we again consider the shining of the developing queen discussed in the previous section. It seems that this shining can be tolerated by the bees only when the old queen is present. Thus, if we remove the old queen during the initial phase of this shining, it is easy to imagine that the counterbalance needed for tolerating this phenomenon is taken away from the bees, and that over several generations this develops into a weakness that the chalkbrood pathogen is able to exploit. But if the bee colony can gain some relief in the harmonious process of liberation by swarming, this weakness does not arise. Even so, the question remains open as to what happens

> *The environment is often held responsible for weaknesses that result in diseases developing. But perhaps a large part of these weaknesses would be removed if we were to try to better understand the essential nature of the bee.*

with the remaining colony that no longer has the counterbalance in the form of the old queen. Here, other natural laws seem to apply about which we do not yet have sufficient knowledge, but if we delay queen removal from the swarming colony until the first swarm cell is capped, this chalkbrood phenomenon does not arise.

However, it is then important that not only is the queen taken, but also part of the colony along with her so that something similar to a natural swarming takes place. This method nevertheless denies the colony the experience of issuing a swarm. Even in this case I wouldn't call it a swarm substitute, rather I see it as an emergency solution which, if viewed long-term and practised repeatedly, weakens the colony in ways that we do not yet fully understand. We know only that our bees become disease prone. The environment is often held responsible for weaknesses that result in diseases developing. But perhaps a large part of these weaknesses would be removed if we were to try to better understand the essential nature of the bee.

Preventing the swarm but swarm cells are not required for breeding queens

The colony again shows that it is in a swarm mood. As already discussed, in this case the whole brood nest is removed from the colony and only the queen with a capped brood frame with bees is given back to it. The free space in the brood box is filled with light-brown combs rather than foundation. The colony with the queen does not yet have bees capable of building comb because the young bees with developed wax glands are in the brood nucleus. It is true that old bees can restore functionality to their wax glands, but this would reduce the population of flying bees and so substantially reduce the honey yield of such a colony.

The brood nest is placed in a nucleus box at the apiary so that the flying bees can fly back. It can of course be divided into two or three nucs, taking care that each is given at least two queen cells. The queens hatching (ecloding) from them will be replaced in the summer by breeder queens. This means that if we are not wanting to practise rejuvenation of colonies with swarm queens or queen replacement, we have to carry out queen breeding or buy queens from a breeder.

Controlling and preventing the swarm urge

Many beekeepers are forced to practise swarm control and prevention methods because of the location of their apiary and/or the time they have available for their beekeeping. Such beekeepers do not usually carry out any natural methods of colony increase, but instead use the various possibilities offered by artificial queen breeding.

For example, imagine an apiary consisting of twenty colonies that are kept in a bee house. The colonies have bred queens. The beekeeper does not want any swarms to issue because they cannot be at the apiary during the day, neither have they the possibility of siting colonies whose queens have still to be mated outside the bee house. The mating in such a large-scale flight of twenty colonies would lead to large losses because queen drifting can be very high. In a situation like this, the beekeeper is forced to prevent swarming.

The simplest method to prevent swarming would be to cut out all queen cells at the various stages of development. By doing this we can suppress the urge to swarm. However, this does not destroy the intention of the colony to swarm. The colony will immediately build queen cells again. If the beekeeper re-inspects the colony a week later, they will again find queen cells at various stages in the same colonies. The beekeeper cuts these out again and after a further week these colonies will possibly have no more queen cells. The beekeeper may be satisfied that they have done the right thing. Only at the honey harvest, and in the behaviour of the colonies, will the beekeeper discover (assuming they are observant enough) that something is not quite right with these colonies. They may ascribe the apparent problem to the queen, as this is the easiest explanation, rather than in the way they have managed their colonies.

What has gone wrong?

This mood recedes if the swarm has issued or, for example, the queen is removed with part of the brood nest. But if *only* queen cells are removed, the swarm mood continues and the colony builds new swarm cells. Although cutting out queen cells eventually causes the colony to abandon its urge to swarm, it is a severe way of doing so. This capitulation not only spoils further industrious foraging for honey, but also results in a pervasive discontent among the bees that can lead to an increased inclination to sting.

If swarming has to be prevented, it would be better to use other options for swarm control so as to maintain harmony in the colonies. Large hives are better for this as they provide more space for colony development at the beginning of swarming time, although changing to bigger hives is only possible with a substantial increase in expenditure, which not all beekeepers can afford. Another possibility is to choose a particularly un-swarmy breed of bee.

If a different bee type is under consideration, you should also think about neighbouring beekeepers. If you are in a region with a pure breed, bringing in a different bee can risk wasting the long time it took to create the pure breed region. If it is known that a neighbouring beekeeper is doing apiary mating, it would be un-neighbourly to endanger their success by introducing your own new bee type. Therefore, when choosing another bee, it is advisable to discuss with neighbouring beekeepers the types of bees that can be imported without question.

But there is another option for controlling and preventing swarming. This involves a deliberate *weakening* of the colony through a careful removal of young bees and brood. This method was first developed using Ernst Perkiewicz's Marburg box, and done properly it can help to regulate colony strength and colony development as a whole.

At the beginning of the swarming mood, bees are removed from the building frame and from a brood comb with open, uncapped brood (eggs and larvae). These bees are brushed with a feather into a box, taking care not to brush the queen in too.

If the colony already has queen cups with eggs, then the bees are removed from the building frame and two brood combs and brushed into the box. Bees should always be taken from combs with open brood as this is where young bees are found who have just developed their brood food glands and are active as feeder or nurse bees. In addition, one, or better still, two combs of freshly capped brood with bees are taken out intact and replaced with light-brown combs. These may very quickly fill with eggs, so the nurse bees can in a few days release their excess brood food.

Freshly capped brood is taken as this will delay the cell area being used for laying again the longest. Freshly capped brood is recognisable from the fact that the brood area is capped outwards from the centre of the comb and as yet open brood is still present in the margins. This is because the queen lays by working in a spiral from the middle to the edges. The oldest eggs, and therefore the first to be capped, are in the middle and the youngest at the edges. You must remember to cut out all queen cups, with or without eggs.

If the colony has gone so far into swarming that there are already queen cells with round larvae, so-called *pitcher-like* cells, or even a *capped cell*, then a radical weakening of the colony is indispensable. There are two options:

- Take the whole brood nest out of the hive and return only the comb with the queen and the bees already on it. The remaining empty space is refilled with light-brown or white fully built combs.

- Look for the queen and remove her. If you remove two or three hatching brood combs you can make a small nuc with them. Hatching brood combs are recognisable by the first bees starting to hatch (eclode) in the centre of a capped comb. As it can now be assumed that the older queen cells will not be so carefully attended to, queen cells of all ages are removed, except cups with eggs or young larvae. The swarming urge now disappears from the colony and these young cells are given the best care.

Now comes the question of what should happen to the bees or combs removed from the weakened colony. If there are weak colonies in the apiary, these combs and/or bees could be used to strengthen them. Another possibility is to create nucs, if there is sufficient material available. Creating nucs is described in detail in the next chapter.

The Marburg box

The Marburg box is a kind of portable nuc box. On one side is a large ventilation mesh that can be covered with a wooden sheet, and below there is an entrance that can be closed off. On the other side, across the entire length, is a large opening covered with a queen excluder and a folding flap or funnel. The bees can be brushed into the funnel in front of the queen excluder and they will then crawl through it into the box; neither queen nor drones are able to get into the box. When the funnel is closed, the whole queen excluder is covered. If young bees from a colony have been brushed into the funnel, it is important to check soon afterwards that the queen has not been brushed in by mistake. If she has, she will be seen crawling on the queen excluder. Only after this check has been carried out should bees be brushed from a further colony.

So that the bees move quickly into the box, it is arranged so that light enters the funnel. This produces a light/dark effect. Because bees aim for the dark, they will quickly leave the funnel and crawl through the excluder into the dark interior of the box. During this procedure, the ventilator mesh should be covered with the thin wooden sheet so that complete darkness prevails inside the box.

If there are many colonies to be treated in this way, it is very useful to insert into the space in the lid of the box any caged queen that may be available. The bees crawling about in the box will quickly find her and hang like a swarm around her. This enables the beekeeper to repeatedly let bees crawl into the box over a period of hours without constantly having to close the funnel flap each time. If no queen is inserted within an hour at most, the young bees will start running around looking for their queen and the light/dark effect will become ineffective. This shows up when the agitated bees crawl out of the box through the excluder and look for the queen in the funnel, thus making it more difficult to weaken additional colonies. The Marburg box is also suitable for making artificial swarms. As it has an entrance, or flight hole, it can be used in the meantime as a nuc box.

The Marburg box method is very time consuming. Additional colonies are always needed into which the removed bees and combs can be introduced. If there are enough of these, then nucs or artificial swarms can be made. The Marburg box method should be used on light/flower or warmth/fruit times.

This method is primarily for beekeepers who do not want any swarming and is therefore not the most bee-centred of methods –

FIGURE 16. Swarm box (left) and Marburg box (right).

although it is a lot more bee-friendly than cutting out queen cells to prevent swarming. If the method is applied properly, however, a very good honey crop can be obtained. The bees remain gentle and colony development is consistently positive. For these reasons this method is still used by many beekeepers.

Whether a beekeeper has swarmy or swarm-resistant bees, the swarm instinct is entirely natural: a bee without a swarm instinct would be unthinkable as a species. If the beekeeper were to put the bees' interests first, then they would allow the colony to swarm every time, but many decide against swarming.

There are numerous reasons for this decision. Since most beekeepers have to go to work, they cannot be in their apiary during weekdays. As the swarm would most likely fly away and the beekeeper would lose not only the swarm but also its honey, the bees are therefore not allowed to swarm. A colony that swarms before the main flow is no longer in a position to produce a good yield of honey. Furthermore, the beekeeper may be concerned that in the course of time the 'swarminess' of their bees will increase through the process of inheritance. The beekeeper may also have expensive breeder queens in their colonies and not want to risk losing them in a swarm. Finally, many beekeepers have too few hives and really do not know where to put the swarm or its parent colony.

Although there is much here for beekeepers to consider, especially in relation to their own circumstances, when it comes to swarm prevention the needs of the bees should be put first whenever possible.

Chapter 6
Colony Regeneration and Propagation

Colony rejuvenation may proceed naturally when, for various reasons, colonies re-queen themselves and swarm, or it can be controlled by giving them a young queen or a nuc containing a young queen. When it occurs naturally, bee colonies have three ways of rejuvenating their queens. We distinguish between swarm queens, supersedure queens and emergency queens.

* A swarm queen appears in a colony when the colony gets into a swarm mood. During this period it may produce as many as two dozen queens. These queens are not all equally cared for, as revealed by the fact that big differences in queen size result. The rejuvenation is successful when the old queen and a portion of the workers issue in a swarm and a young queen hatches in the residual colony, thus helping it rejuvenate.

* A supersedure queen arises when, for various reasons, a colony is not satisfied with its queen. In most cases of

supersedure, colonies make two or three queens. The impression is gained that colonies place great value on the good care of these queens. In supersedure colonies, there arises the only situation in which two queens coexist for a long period without fighting, each with her own brood nest. The old queen will at some point fly out and die. Supersedure occurs mostly in summer or autumn.

❋ In exceptional circumstances, raising emergency queens is the only way that bees can obtain a new queen. Bees are able to raise a queen from a worker larva by changing the shape of the cell and the brood food it receives. This situation may arise when, for whatever reason, the colony has lost its queen. Without a queen to lay more eggs, the colony would die out in a few weeks when the last of the workers reaches the end of their lives. To counter this, the bees look for worker larvae and partly chew away the walls of the cells containing them. These shortened hexagonal cells are then rounded off. This has to be done because queens are raised only in round cells. The larvae are fed with royal jelly and emergency queens are produced.

However, emergency queens do not appear to be regarded by colonies as queens in the fullest sense, as they are often superseded at the latest in the following year. The reason generally given for this is that the colony has selected from the older larvae in order to get a queen more quickly, but in so doing it gets one that is not as good as a queen raised from a really young larva. The larval age affects the number of ovarioles in the ovary: the older the larva used to raise a queen, the smaller the number of ovarioles developed in the ovary. This explanation is quite illuminating. Except we need to ask why then, if this is the case, do bees not raise only the youngest larvae? The real reason is actually quite different.

Examination of swarm or supersedure cells of queens shows that they are round all the way down, from the cap to the cell floor: from the outset, the egg is laid in a round cell. This round cell is then extended in the course of queen development to form a round sac that finally takes on the shape of a grape. With emergency queens, however, the egg is laid in a hexagonal cell because a worker bee was originally intended to develop from this egg. The larva hatches after three days and after two or three days more this larva is chosen by the workers to become a queen. This means that for up to six days the future queen develops in a hexagonal cell, not a round one. If we accept that different shapes have different formative influences, we will be able to understand that a queen that has developed from the start in a round cell has received different influences from one that has spent about a third of its development in a hexagonal cell. Added to this is the fact that the bees can only round-off the walls of the worker cell and not the entire cell floor, which retains its hexagonal shape. The influence of formative forces is discussed in more detail in Chapters 8 and 9, which look at comb construction and honey.

A further option for queen rejuvenation involves breeder queens. This is an artificially conducted and controlled method of raising as many queens as required, and is discussed in more detail in the following chapter on breeding queens.

Natural increase in colony numbers

An increase in the number of colonies can be allowed to happen naturally or it can be influenced by the beekeeper. Natural colony number increase happens exclusively through swarming, and how many times a colony divides itself by swarming depends totally on the colony. There are very swarmy colonies that, after the prime swarm has departed (the one containing the old queen), divide further and issue up to three more casts (secondary swarms or after-swarms), sometimes more in extreme cases. Less swarmy colonies are often satisfied with issuing just the prime swarm. Thus, if a natural increase in colony numbers is desired, it is possible only through swarming.

As already described in Chapter 5 on the swarm urge, this method of increasing colony numbers is no longer normally used, with the exception of beekeeping that has specialised in intensive swarming reproduction. However, what should we do if we get swarms and how should they be treated? Those who are used to dealing with swarms know how sensitively bees can react. If a serious mistake is made in handling the swarm, the bees reward the keeper by unceremoniously flying away again. It is therefore necessary in this context that, as with controlling the

FIGURE 17 (left). This swarm did not accept the hive and settled in a tree.
FIGURE 18 (right). The swarm later settled down on the front of the hive.

swarm urge, we try to understand the swarm itself in order to treat it in a way that is appropriate.

We will assume that our colony has swarmed and the swarm is hanging in a tree. *Swarms do not usually land where beekeepers want them to. Normally a swarm clusters where the queen has settled, or where the scout bees have found an appropriate place.* If you have not yet gained any experience with swarms, you should note each time exactly where the swarm is hanging, so that you can hang up a swarm catcher at this location in order to greatly simplify the 'knocking-in' or taking of the swarm.

The swarm, now hanging in a tree, is lightly sprayed with water

FIGURE 19. *A swarm in the eaves of a roof.*

so as to calm it and somewhat contract the cluster. For knocking-in the swarm, it is best to use a swarm basket (also known as a skep) or a swarm box. *Weigh the empty skep or swarm box*, so you can later work out the weight of the swarm. Once the swarm is taken, the basket or box is placed beneath where the swarm was found hanging and left until evening. The swarm will stay in the box only if the queen has also been taken into the box. Therefore, the swarm catching box should be left under the place where the swarm was knocked (or shaken) into it. If the bees are still in the box in the evening, the queen will be there too. If the queen is successfully taken into the box, the bees will gather around the entrance and start fanning and emitting pheromone from their Nasonov scent glands. In doing so they evaginate the gland and 'blow' its scent into the air. This enables bees still flying around to find the entrance and join the swarm in the box. If the bees come out again, say within about half an hour, this means that the queen has not been taken into the box. The swarm then reforms on the place where the queen has landed and it must once again be knocked into the box. In the evening, the box entrance is closed and the box placed in a quiet, dry, dark cellar. Ventilation is provided by fully opening the ventilation mesh. With a basket, or skep, a piece of large-pore linen fabric is stretched over the top and the skep must be so positioned that enough air can pass through the fabric. The swarm remains in darkness for 24 hours, during which time it should become *mature* or *ripe*. This is indicated by its hanging in a harmoniously formed cluster.

FIGURE 20. *Here a swarm is knocked into the box with a simple thump on the shed roof.*

To conduct the further handling of the swarm in accordance with natural swarming, it is placed at a new site. A hive is prepared, its inside scraped with a hive tool and then scorched. This gets rid of old wax and dirt residues, and the scorching kills pathogens as well as helping the hive take on a bee-friendly smell through melting the propolis – a resinous mixture produced by bees to fill narrow gaps in the hive. If a swarm is put in a dirty, musty-smelling hive, there is a very real possibility that the bees will dislike it so much that they will swarm again as soon as possible.

The new hive should consist of two superimposed boxes. In the upper box are hung either light-coloured combs, foundation, or, if the swarm is to be left to build freely just as swarms naturally prefer, building frames set up as discussed in Chapter 4 on swarming. In order to give the swarm enough combs, it is

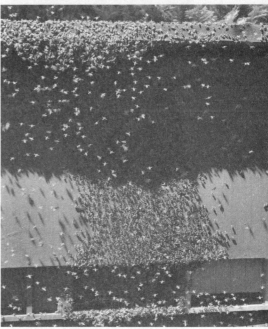

FIGURE 21 (top). In this case the swarm was just left in its box on the roof until evening.

FIGURE 22 (bottom). A very large swarm, that had initially hung under the roof of the shed and then marched in through the hive entrance.

weighed and for every 500 g (1 lb) of bee weight, two combs or frames are given. The bottom box remains empty. This bottom box, into which we knock the swarm, gives the swarm enough space to hang under the combs or frames, leaving it to occupy the combs when it wants to later on. Swarms have repeatedly shown that they like to be able to hang and not be separated into parts by combs or foundation. When, after a few days, the swarm has moved into the top box, the beekeeper transfers the occupied frames into the empty box below. I suggest that this job is best done at earth/root times to promote comb building.

If at the time of taking the swarm there is not enough honey forage naturally available, then on the evening of the second day after taking the swarm the beekeeper provides feed in the form of honey syrup or sugar syrup. This is given so that the bees do not starve and to stimulate the building urge. This feed should not be given at the start, because swarms with sufficient stores in their crops generally abscond. If fed right at the beginning, the risk of absconding from a new hive is very high. As the swarm uses a lot of energy for building totally new comb, it is advisable to give it bee tea as recommended for winter feeding (see Chapter 10). Once the swarm has built the combs and has progressed well with raising brood – which is to say, it has established a good brood nest – it can be given a honey super or even a second brood box as soon as the forage flow allows. Under favourable conditions a swarm can bring in a nice harvest.

The beekeeper should next decide what is to happen with the swarmed colony. If they are a heather beekeeper – that is, a late-forage beekeeper – they will gladly accept more swarms. If they want to get as many new colonies as possible, they will likewise wait for more swarms. But the swarmed colony may react very differently from the way the beekeeper wants. For example, if the beekeeper has bees that are satisfied with a prime swarm, then they will not get any more swarms from them, even if that colony has raised a dozen swarm cells. How can a beekeeper tell if they have such a colony? The answer, with a little arithmetic.

The beekeeper knows that the swarm issued on the day *after* the capping of the first swarm cell, that is, on the tenth day of the cell's development. The queen then emerges from this cell five to six days after the swarm has issued. If the beekeeper goes to the apiary on this day, they will hear a softly repeated sound, especially in the evening when the surroundings are quiet. It sounds as if someone is making a high-pitched tooting sound at short intervals. This is indeed called tooting. It is produced by the first swarm queen to emerge. If you are lucky enough to find her in the opened hive, she will be crawling across a comb and pressing her body on the comb, thus making it vibrate strongly so as to emit this tooting sound. The tooting seems to be a call aimed at summoning an answer from the other virgin queens, who are ready to emerge from still-capped cells. Now the queen goes to the cell from which comes an answer that sounds like the quacking of a frog. She bites open the cell in its upper third and stings her rival. This of course seems somewhat gruesome, but it is just nature's way when a colony does not want to issue more swarms.

Knowing that bees react to cosmic rhythms raises the question of whether swarming is also dependent on certain constellations. Based on our various experiences, we assumed that there is such a dependence. However, our observations must continue further, especially as this assumption was reinforced by an event over which many a beekeeper would have torn out their hair.

One of our apiaries with twenty colonies was of necessity neglected one year precisely at swarm time. It could not be inspected for about three weeks. When the inspection once again became possible, we made an enormously interesting observation, which pleased us greatly despite the damage caused.

Of the twenty colonies, eighteen had issued prime swarms. This was clearly a big loss to us because these eighteen swarms had flown away. Of the eighteen swarmed colonies, fifteen were no longer inclined to swarm and all fifteen colonies were tooting – we needed only to lift the hive cover to immediately hear the queen. We had never experienced anything like that until then. This fact showed us that all these colonies came into swarm mood simultaneously and that this was without doubt attributable to a constellation, about which we had to make further investigations.

Prime swarm at the site of the parent colony

If there is a main flow at the time of swarming that we want to fully exploit, if we want to achieve a good harvest from the swarming colony, then we must employ the method known as *prime swarm at the site of the parent colony*. However, we need to be clear that this method does not accord with the nature of the bees; the swarm does not want to return to the site of the parent colony.

The swarm is treated as described in the previous section: it is taken, placed in a cellar and *processed* on the next day. Meanwhile, the swarmed parent colony is moved to another site in the apiary. At the vacated site we put our cleaned, empty hive into which we shake the swarm. Now all flying or field bees from the swarmed parent colony can fly back the next day and further strengthen the swarm. But as we want to make use of the current nectar flow, after two or three days we give the swarm a honey super from the swarmed parent colony. As the swarm may have acquired a somewhat different smell from the new hive, it and the super are sprayed with thyme water. The uniting works best this way. Anyone who feels unsure about this may insert a sheet of newspaper punctured with fifteen to twenty small holes (use a nail or ballpoint pen) between the swarm and the super. The bees make contact with each other through the holes. If all goes well, then they chew away the newspaper. This leads to a harmonious uniting between the swarm bees and those in the honey super. We have found that the swarm can bring in a good honey harvest with this procedure.

Some beekeepers may not want the swarm that is hanging in the tree. In this case they sieve out the queen with the help of a sieving box and let the bees fly back to the parent colony. But this only works if the swarm has not yet hung overnight. The bees will only fly back when the queen is sieved out of the swarm or caught on the day on which the swarm issued. If it has hung overnight on the tree, then the swarm has already inwardly separated itself from the parent colony. If it is de-queened now, it will not find its way back to the parent colony. Among beekeepers it is said that when a swarm has flown it forgets the place from which it has issued.

A swarm that has been de-queened in time, can still only be returned to the parent colony through the hive entrance. To this end a platform of cardboard or wood is set up directly in front of the parent colony and a good dessertspoonful of bees is placed on the platform to see if the parent-colony bees will accept the swarmed bees. This occurs when the guard bees of the parent colony examine the swarm bees by walking around them and stroking them with their antennae, and the swarm bees then stand and begin scenting from their Nasonov glands. All the bees may now gradually be emptied onto this platform so that the swarm can crawl back into the parent colony. Even this method cannot unequivocally be described as bee-friendly, because it easily happens that even though the colony is lacking the old queen, the swarm mood starts again. The swarming bees really wanted to find a new home and new surroundings, and this was thwarted by our intervention.

*Figures 23 and 24. Left: Cardboard running-in platform.
Right: Running-in with an inverted frame rack.*

Artificial colony increase

If the beekeeper wants to control the increase in the number of
colonies, they have two possibilities. Either they use the swarmed
parent colony with swarm cells for making nucs, or they breed or
buy queens to use for making up nucs.

Making nucs (colony reproduction) with swarmed colonies is
a simple method and one acceptable to the bees. Because we want
to support the process of colony enlargement by making nucs,
this work is done at light/flower times. If this is not possible, we
may choose warmth/fruit times. However, the first nuc inspection
should then be done at light/flower times.

The brood combs from a swarmed colony should be grouped
in at least threes, preferably more. Each group should have
several swarm cells at various stages of development, from cups
to capped cells. In order to reduce any desire to swarm still
present, all capped swarm cells are cut out. However, these cells

should still be cut out even if no importance is attached to the still existing swarming urge. It was previously indicated that a swarming colony does not care for all queen cells equally. In order then to get optimally developed queen cells, the capped cells are cut out. As it must now form itself into a new colony organism, the prepared nuc ensures that the remaining uncapped queen cells are well cared for. To be totally sure that only well-developed queens arise, after four or five days we can once again cut out the oldest cells. But the nuc must be left with two or three queen cells, because it can happen that a cell does not hatch (eclode).

The nuc was put together with brood combs containing queen cells. If there are still combs with honey or pollen stores available, they are shared amongst the groups of brood comb. The prepared nucs must comprise at least six combs covered with bees, and it is necessary to ensure there are sufficient stores present as the nuc will lose its flying bees. If there are not enough combs of stores in the swarmed colony, dry combs of feed stores from your comb store should be given. There is a risk of robbing if these combs are not really dry.

For nucs we use either nuc boxes that should have space for at least eight frames, or a single box from a top-access or magazine hive. These boxes are equipped with a lid and floor, so that a complete hive is assembled. The finished nuc is placed at the side of the apiary. The flying bees fly back to the parent colony and support nectar foraging there. A water supply is also necessary, because the nuc has large areas of brood that need feeding and keeping warm, as well as having its temperature precisely regulated. However, because the water-carrying bees are among the flying bees that the nuc has given up to the parent colony, a comb filled with water is placed to the side of the other combs. The water comb must not be hung immediately next to a brood comb, but should be right on the edge of the colony so the brood is not harmed by its cooling effects. The bees can now use this water comb to meet their water requirement.

After a few days, the queen will emerge in the nuc. She is of course not yet mated and still awaits her mating flight. To facilitate the queen's orientation and to better enable her return to the nuc, something conspicuous is erected or placed in front of the nuc and should stay there until the queen is demonstrably mated. It is for this *orientation reason* that the nuc is placed at the edge of the apiary, because losses due to possible queen drifting would otherwise be too high. For example, large bee houses with close-packed hive frontages often reckon with losses of 50 per cent due to queen drifting. This can be avoided when a nuc whose queen is still to be mated is situated alone and equipped with good orientation landmarks.

If the cutting-out of queen cells from the nuc has been performed, and enough feed given, the nuc may be left alone for three weeks. In this time, the queen eclodes, is mated and starts laying. If such a nuc is disturbed too soon, it can easily happen that the queen is balled and stung by the bees. There is still no plausible explanation for this phenomenon, we know only that it does not happen when

FIGURE 25. *Orientation landmarks for the queen. There is a branch in front of the second hive from the left. The box stack and the closed parasol are also good landmarks.*

the queen has already started laying. One possible reason for it might be that a very sensitive situation has arisen in the parent colony from which the swarm has just issued together with the old queen. If the parent colony is then completely taken apart for making nucs, the bees become all the more sensitive. We should therefore be more aware of the inner life of such a nuc.

A bee colony is an organism and this organism is less sensitive to outer influences when it is healthy and intact. It contains a mated queen as well as drones and workers. The comb is likewise part of such an organism. If one of these components is missing or is incomplete, the organism becomes delicate and the colony's harmony is threatened. If we make a nuc in the way I've described, it too will turn into an organism, but this formative process is only completed when the nuc has a mated and laying queen. If we inspect the nuc too early, this will disturb its development into a harmonious organism to the point that the bees will ball an incomplete, unmated queen. In some cases, this could mean that she is killed.

Major interventions in the bee colony organism should therefore be undertaken only with utmost consideration for the essential nature of the bee. In the long run, beekeepers will only be happy with bees when they try to understand them and do not violate them.

Major interventions in the bee colony organism should therefore be undertaken only with utmost consideration for the essential nature of the bee. In the long run, beekeepers will only be happy with bees when they try to understand them and do not violate them.

Colony reproduction with bred queens

If queen breeding – often referred to as 'artificial queen breeding' – is to be undertaken, beekeepers have various options available for obtaining breeding material. They can obtain eggs, which is the most difficult way, graft larvae, or use the cell punch method. As starting material, we have worker larvae which we know can be raised as queens through changing the cell shape and the brood food.

The worker larvae should be between one and two days old. We have already discussed the effect of larval age on the resulting queens. The larvae are lifted out of worker cells with the grafting tool – the best for this being the Swiss-style grafting tool – and placed in a queen cup cast by the beekeeper. These cups are then transferred to the care of bees that are willing to undertake it, usually bees in queenless colonies, or in specially prepared, queen-right colonies.

Another possibility is the cell punch method. A piece of comb containing the youngest larvae is cut out of a brood comb. Using a sharp knife that has been drawn across a stone to give it fine teeth, cut the cells short to half their depth. Now using the cell punch, punch out whole cells with their individual larvae. The punched-out cells are clamped in so-called clamp plugs (*Klemmstopfen*) and likewise given to colonies ready to look after their further development. This transplanting of larvae or punching out of cells should be done when Venus and the moon are in a light/flower constellation. These times give the best queen reproduction results. Further development up to the mated queen is dealt with in the following chapter on queen breeding.

If the mated queen, or a small mating colony is ready, nuc making or colony reproduction can proceed with bred queens. There are many options for colony reproduction. Below, we describe four of them. We select light/flower times for their cosmic influence; if necessary, we can switch to warmth/fruit times, but further work on the colonies should take place at light/flower times.

Various options for colony reproduction

The beekeeper has a number of options to choose from when it comes to colony reproduction:

* Artificial swarm

* Nuc with capped brood and single queen

* Nuc with capped brood and small mating colony

* Nuc with brood at all stages and an individual queen or small mating colony

Artificial swarm

As the term already suggests, this involves making a swarm artificially. The necessary equipment comprises kitchen scales, a swarm box, a spray bottle and a large sheet metal funnel that is put on the swarm box so the bees can be brushed into it. The swarm box should be equipped with a device into which a queen cage can be inserted for queening the swarm. This should be done at light/flower times, in contrast to nuc making, which is done at warmth/fruit times.

FIGURE 26. Swarm box with funnel.

A supply of bees can be obtained from honey supers. For example, if you have ten colonies, you can make three or four artificial swarms without excessively weakening the colonies. Enough bees are brushed into the swarm box until it registers about 2 kg (4½ lb) of bees. If bees are brushed together from several colonies, they tolerate each other because they are removed from their colony contexts and have no queen. Of course, when brushing bees together, take care not to brush a queen into the swarm box. Keep the funnel wetted with the spray bottle so the bees slide easily through it and do not cling onto it. Now quickly jolt the swarm box to knock down any bees hanging on the bottom part of the funnel, then remove the funnel and close the swarm box. Finally, put the cage with the mated queen in the holder provided, and place the artificial swarm in a cool, dark, quiet room.

In order to avoid any forage loss through removing bees from supers, it is advisable to make artificial swarms during the period after the main flow. Unmated queens should not generally be used for artificial swarm or nuc making as they are very rarely accepted. The queen only becomes a completely mature queen after she has been mated. The bees clearly seem to recognise this and tend to ball and sting unmated queens. An exception to this is in small mating colonies, although even here particular conditions must still be observed. The artificial swarm remains in darkness for 48 hours. This is the time necessary for allowing the brushed-together bees to form themselves into an organism.

In order for an artificial swarm to arise from brushed-together bees, it has to undergo a maturing process. The beekeeper now has the possibility of observing how an immature swarm becomes a mature one. If the bees are studied through the ventilation mesh, they are seen hanging right around the inside of the swarm box. To promote the 'growing together' of the bees into an organism, in the evening the bees are given 1 litre (1 quart) of feed in a top-feeder. Feeding has repeatedly shown to positively influence the process of the bees uniting. Twenty-four hours after preparing the artificial swarm, it may already be clear whether they have accepted the queen given in the closed queen cage. If they have drawn together into a cluster, it shows that they want to form themselves into an organism.

Now open the holder into which the queen cage was inserted, take out the cage and observe how the bees around the cage behave. If they are trying aggressively to get their heads through the mesh it shows they have not yet fully accepted the queen. In this case the queen should not be released yet. You need to check again in 12 hours to see if the queen has been accepted. If the bees are still rejecting the queen after 48 hours of darkness, you can assume that this is either attributable to the queen herself, or even to another queen having been accidentally brushed into the swarm box with the bees. In the latter case the bees have to be sieved through a sieve box and the queen removed. After appropriate checks on the colonies that provided the bees, she should be reintroduced in a closed cage into whichever colony is queenless. Because in the interim this colony may have made emergency cells, the whole brood nest needs checking. After 24 hours, the queen can be released by using a soft candy plug in the queen cage. Over the course of the next few hours the bees will chew through the plug to free the queen.

If the rejection of the queen by the artificial swarm bees is attributable to the queen herself, the artificial swarm needs supplying with another queen. If, when the cage is checked, the bees crawl around on the cage and can be dislodged by gently tapping with a finger, the queen has been accepted. The cage plug can now be removed and re-closed with soft candy and the cage replaced in the queening slot. The artificial swarm is given another litre of feed because the feeder will most likely have been emptied in the meantime. As the queen can now move freely in the artificial swarm, the aim of forming an organism is achieved more quickly. When 48 hours of darkness have elapsed, knock down the artificial swarm as described previously for natural swarms. Knocking down the artificial swarm reveals whether it has really become an organic unit. If you examine the underside of the swarm box lid where the swarm was hanging, you will see small natural combs there. When the swarm has matured or grown into a harmonious organism, these combs will be built correctly and consist exclusively of worker cells. If the second litre of feed was not given, this construction might not occur.

Of course, you might now ask why, in contrast to nucs, artificial swarms are made at warmth/fruit times and not at light/flower times. In the experiments we did with swarms and artificial swarms, we always noticed that swarms knocked down or brushed in at earth/root times and given foundation or building frames, built better than swarms taken at other times. Artificial swarms should spend 48 hours in darkness, so if one is shaken or brushed in at an earth/root time, it must as a consequence be created at a warmth/fruit time.

Virgo's influence lasts for about four days, so there would be the possibility of carrying out both the creating *and* brushing in of an artificial swarm at earth/root times. But as we have found that earth/root times are generally not very favourable for working with colonies, and we have to open several when creating artificial swarms, we choose the more favourable warmth/fruit times. This way allows an artificial swarm to mature during the transition between warmth/fruit times and earth/root times and be knocked-in at earth/root times, thereby greatly supporting the urge to build comb. In view of the positive effects of warmth/fruit times when the artificial swarm is created, we might assume that the positive influences also affect the building urge. Our experiments have time and again shown that when the organism is disturbed, the momentary cosmic constellations find expression in the bee colony.

It is clear from the artificial swarm's maturing process that an artificial swarm can consist of bees from different colonies. We give it a queen, provide feed and it gradually develops into an organism. The process culminates when the swarm begins to build worker cells. We can now say that the organism has been created, although for full maturity it still lacks brood. Now the beekeeper fetches the artificial swarm at earth/root times and knocks or shakes it into a hive. The result of this knocking-in is that the artificial swarm totally disintegrates and must now once again reorganise itself into the organic shape of the swarm cluster. To facilitate this the beekeeper gives it space under the frames that are to be built-up with comb. Thus, the artificial swarm forms itself into an organism again under the influence of earth/root times. The climax of this is that the newly formed swarm cluster moves up into the comb or frame assembly and starts building comb and raising brood.

Thus, the artificial swarm is formed at warmth/fruit times so that the colonies from which the bees were removed will be disturbed on favourable days. In contrast, the knocking-in of the artificial swarm occurs at earth/root times, because that promotes the building impulse. Further inspections should again be done at light/flower times. Since there is not usually much forage available at the time of making artificial swarms, in the evening following the knocking-in, it is advisable to feed the artificial swarm 5 litres (5 quarts) of liquid feed, preferably honey syrup, in order to support comb building and brood production as well as to provide a feed foundation.

Nuc with capped brood and single queen

If we do not breed queens ourselves, we are dependent on bought-in queens. We receive a caged queen with about fifty bees. It is desirable to have this queen available as early as June so that she is surrounded by her own bees until the feeding, and already in the first year has her own colony to build up. The preparations and the nuc making itself are done at light/flower times. The preparation for nuc making consists of hanging three or four brood combs behind or over the queen excluder of production colonies so that the queen cannot continue to lay eggs on them. After nine days, at the next light/flower time, all cells are capped and the nuc can be made. Before starting nuc making, the hive into which the nuc will be placed should be prepared as already discussed. Some beekeepers still use rather small nuc boxes into which a maximum of five frames of comb will fit. It is most certainly better to use hives into which at least eight combs can be inserted. Honey supers equipped with roofs and floors serve well for this. The large space of ten or twelve frames is then divided with a straw divider or a division board. Nuc making comes after these preparations.

The brood frames previously separated from the queen by the queen excluder are now placed on the comb/frame rack. If beekeepers are slow in this work with the bees, they should cover the brood combs on the comb rack with a tight-weave linen cloth so that the combs do not cool down. Four brood combs with bees are now hung in a nuc box. We then give a dry, light-brown comb to serve as a front comb, and we hang a dry comb of honey as an end comb at the back of the brood combs. (This assumes combs are hung parallel to the front of the hive, known as 'warm-way' orientation. If combs are hung perpendicular to the front of the hive this is known as 'cold-way'.) The combs given should be dry because, if such work is undertaken on days with no nectar flow, feed combs wet with honey or feed can easily trigger robbery. Now

FIGURE 27. *Kirchhain Mating Nuc Box (KMNB). The angle of the side walls means the bees do not stick combs to them, which makes comb inspection very easy.*

FIGURE 28. *KMNB comb. The box holds six small combs on which the colony can develop for several weeks.*

close the nuc box. So that the nuc can defend itself in the event of robbery, the entrance should not be too large. When the bees realise that they are queenless they are given a queen. The queenless condition is observable after thirty minutes to an hour, by bees crawling around looking for something on the nuc's front. The queen is then released from the cage in which she was received from the breeder, and transferred to a queen introduction cage that fits between the combs. In doing so, take care not to include workers in the introduction cage. The further procedure is exactly as described for the artificial swarm.

Next evening, check whether the nuc has accepted the queen. Then on the following day free her by means of a soft candy or fondant plug. When introducing the queen in a closed cage, we provide a hand-sized lump of solid fondant. The nuc should not be given liquid feed as this leads easily to robbing. It should always be kept in mind that the brood combs used for making the nuc also hold flying bees that will fly back to their parent colony. If these bees find combs wet with honey in the nuc, or they are given liquid feed, they will fly back to their colony and report that they have found a source of forage. The nuc will now not be able to defend itself from scouts that fly to it and, as a rule, will be lost, unless it is taken as soon as possible to another apiary. When doing this, always work quickly and carefully. The danger of robbery is reduced, or even non-existent, if it is possible to make nucs during foraging when colonies are distracted by a prevailing nectar flow.

Light/flower times are used for this type of nuc making. We simply put together capped brood combs, a dry, empty, light-brown comb, and a dry comb containing honey. After the nuc realises it is queenless, a mated queen is introduced in a closed cage. The colony is fed with fondant and 24 hours later the queen is released using a fondant plug.

Nuc with capped brood and small mating colony

In this option, the process of creating a nuc is the same as in the previous example; only when the queen is introduced along with the small mating colony is the further development of the nuc a little different.

If we give the nuc not only a single queen but also the small, mating nucleus colony, then we can count on a greater degree of harmony in the resulting colony. This is because the queen is already surrounded by an organism that she brings with her into the newly developing organism of the nuc. It is not the queen that is evaluated by the nuc bees from the capped brood, rather it is the organism of the small mating colony that decides whether or not it wants to unite with the nuc. If, for example, a queen cell or an unmated queen has got into the nuc unnoticed by the beekeeper, the small mating colony will not unite with the nuc. The behaviour of the mating colony in this regard is a helpful indication that something like this might have occurred, whereas when only a single queen is introduced, it is not always clear why the bees will not accept her. If the beekeeper then becomes impatient and tries to introduce the queen using a candy plug, she is often killed by the bees.

Introducing a small mating colony to the nuc proceeds in the following way. The nuc is already assembled and the bees are aware that they are without a queen. The entrance of the small mating nucleus colony is now closed and the glass sheet or cover is removed. A sheet of newspaper, pierced with between ten to twelve small holes using a nail or a ballpoint pen, is placed over the opening and held in place with the help of rubber rings. The first exchange of scents takes place through these small holes, and they

FIGURE 29. *KMNB fitted with newspaper and pierced with small holes to facilitate uniting.*

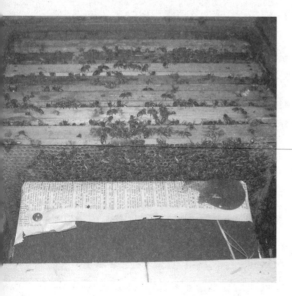

FIGURE 30. *The mating nuc placed with 1 cm (½ in) gap between it and the combs.*

provide ideal starting points for chewing away the paper. The small mating colony is placed with the newspaper side facing the rearmost comb of the nuc. There should be a gap of about 1 cm (½ in) between the stretched newspaper and the combs. The nuc is given some fondant which in turn should promote uniting. If the nuc is genuinely queenless, the newspaper will be gnawed through overnight and the small mating colony will unite with the nuc. An organism now arises from this assembled nuc. After a few days, the queen leaves the small mating colony and begins to lay in the nuc's brood nest.

Nuc with brood at all stages and an individual queen or small mating colony

Making nucs with brood combs containing brood at all stages is increasingly popular, because it is thought that having brood at all stages better preserves the harmony of the nuc. This method is harder to implement because making nucs with small mating colonies and open brood is very difficult. It is advisable, therefore, to use this method only when you have mastered making nucs with capped brood.

If you want to make several nucs, it is best to restrict in advance the movement of the queens of the colonies out of which the brood combs will be taken. For example, this means looking for the queens on the previous day and either restricting them in a comb box that fully encloses the comb and has queen excluders on the large sides, or in a hive chamber using a queen excluder. This makes it possible to work quickly on the day of nuc making without having to keep the brood combs on the comb rack for an excessively long time while looking for the queens. If necessary, the queen

FIGURE 31. *Shredded paper at the hive entrance indicates a successful uniting of the two colonies.*

FIGURE 32. *The newspaper gnawed through during uniting.*

can be caged on the comb itself in a cage that restricts her travel, but still allows the workers to pass through its perforated walls so they can continue to care for their queen. However, restricting the queen in a cage in this way has the disadvantage that she can no longer proceed with her laying work in a harmonious way, and thus a disturbance of the colony organism will very likely occur.

Again, light/flower times are chosen for this nuc making as they favour colony development. At least four combs of brood and bees are used for each nuc as well as a dry, light-brown comb and a dry comb of honey. These are placed in nuc boxes. As with nucs with capped brood, when the bees show that they are queenless, give them a queen in a closed cage and a hand-sized lump of fondant. Great care should be taken to hang the queen cage in the centre of the nuc, so that the bees can quickly find her. This is vital in this method, because there are eggs and larvae on the combs. If the bees do not find the queen quickly enough, they will make emergency queens from some larvae. This in turn can lead to the queen not being accepted. It is clear from this why there is an increased risk of queen loss with this particular type of nuc production. After 24 hours, if the queen is accepted, she can be released using the fondant plug method.

To be certain that no emergency queen cells have been made, at the next warmth/fruit time carry out a brief inspection of the brood combs. Rarely does a nuc make emergency queen cells, but if it does and if these are not removed, then the possibility arises that the nuc has decided in favour of the emergency queen rather than the queen that was introduced. The result of this conflict is considerable disharmony in the colony.

If you want to introduce a queen with a small mating colony to a nuc containing brood of all stages, unavoidable difficulties will arise. As has already been described, the small nucleus colony is inserted behind the protection of a sheet of newspaper. As the bees normally chew through the newspaper barrier only after some hours, in the intervening time the nuc bees make emergency queen cells. If this happens, the mating colony will not chew through the newspaper to unite with the bees in the nuc. At the next inspection we would be able to remove the mating colony again and find in the nuc emergency queen cells about to hatch. To eliminate this possibility, a brood inspection has to be carried out at the next warmth/fruit time to remove the emergency queen cells that will certainly be present. Viewed from the point of view of the extra work involved, this is really not a problem for the beekeeper. But for the nuc it will be a shock, because it has already completely adjusted to the newly developing emergency queens. Anyone who thinks carefully about this will decide to use small mating colonies for nuc making only when the nuc already has capped brood combs.

Making nucs with brood combs at all stages is possible if you want to introduce single queens. However, you must take into account that emergency queen cells will be made and will have to be removed. Introducing a queen with a mating nuc is not advisable, because the nuc will inevitably make emergency queen cells and very certainly focus on these, only to suffer damage when the cells are removed from it too late.

Chapter 7
Breeding Queen Bees

Beekeepers who do not wish to increase their colonies through swarming or are unable to do so, are dependent on bred queens. Whenever mention is made of bred queens we are reminded of the term 'artificial queen breeding'. This term is misleading as well as annoying, because people automatically think of something mechanically made or brought about by chemical alteration. Fundamentally, queen breeding is a way of reproducing queens that is initiated, controlled and monitored by people. The term 'artificial' refers to the fact that people can make as many queens as they want from a colony's worker larvae. This used to be the topic of many a heated discussion, but is no longer so today. It has long been obvious that a good beekeeper works as far as possible only with bred queens. It means a lot to beekeepers that they can show as many colonies with a thoroughbred pedigree as possible, as evidenced by their breeding cards. This gives the beekeeper a feeling of success, just as the herdbook does for the cattle breeder. But should we put bee breeding on the same level as that of cattle, horses or pigs?

The very nature of their mating shows us that considerable differences exist. Mating in cattle or other domestic animals requires a female and a male animal. With bees it requires a female and many – over a dozen – male creatures. We can have control over the mating of our domestic animals, but we cannot with bees. The queen flies to mate in drone congregation areas situated in the air at certain places in the landscape. Over the course of several flights the queen will mate with on average between twelve to fourteen drones, although it may be more than this. These flights are called mating flights. With a domestic animal, mating must be repeated before each offspring is produced. With bees, the sperm from the drones with which the queen mated in her youth lasts for her whole life. These differences in mating should show people that from the point of view of breeding, two totally different routes have to be taken.

Many mistakes made by bee breeders in the past have been recognised, and now, with race foremost in their minds, bee breeders are trying to breed back the bees currently available in order to arrive at the racial archetype once again. However, this 'breeding back' is primarily aimed at gaining new breeding material. The firm view at present is that weaknesses in bees can be combatted only by crossing various lines or races in order to produce more resistant and stronger bees. The bees have certainly shown us that this is possible. For example, if queens are bred from thoroughbred queens without taking into account the drones, the bees that result from this often have little in common with the characteristics of the breeding colony, but are nevertheless very robust and productive. For many years people have tried to make meaningful use of this effect. Only the future will show us whether this is indeed the right way.

Those doing queen breeding always need to be aware that drones produced from bred queens can mate with unmated queens from any apiary site within a 15 km (9 mile) radius. With this in mind it is easy to understand why beekeepers, who may have spent many years establishing a thoroughbred region, are suddenly up in arms when another beekeeper, for whatever reason, begins to keep a totally different race of bees on the edge of that region. Infiltration by the alien drones is certain and significantly endangers the maintenance of the thoroughbred region.

This example shows that, unlike with cattle, we cannot leave it up to each person to decide what race they keep. We have to appeal to the solidarity of all beekeepers so that no other race is kept out of purely selfish concerns, but instead there is a return to a race of bees that can become native to its environment through breeding and selection on a large scale.

Looking back to the time around the beginning of the twentieth century, it is clear that the diversity of races and crosses that we see among bee populations now was not present then. Across the various landscapes, nature had emphasised particular bee types that could be kept in their respective landscape without being seriously damaged by foreign breeds. This keeping of different races was often protected by, for example, large areas of water such as the Baltic Sea or the Mediterranean Sea, and by mountain ranges such as the Alps, over which swarms cannot normally fly, thereby avoiding the mixing of races.

The fact that today we are to some extent working with races totally alien to the region has become possible only through queen breeding. For example, the drones from an Italian queen brought to Berlin and used for breeding 'contaminate' an area of large radius. This results in undesirable crossing with the local bees. Perhaps the same was done in Frankfurt and Munich, which contributed to the fact that the local bees could never again be reproduced by swarming, but were maintained by breeding and selection. When increasing numbers of beekeepers then realised that queen breeding is a very simple task, they brought in queens from various holiday destinations to see if they would produce better colonies and bigger harvests. At least in the first half of the twentieth century, this was the way that the foundation was laid for beekeeping with many races.

In the meantime, the local northern bee (*Apis mellifera mellifera*) of former times is no longer to be found in Germany. It has been suppressed by the breeding of other races. Meanwhile, some beekeepers have realised that if they want to return to robust, healthy bees, this is only achievable via a native bee, and not through constant arbitrary crosses. It is interesting to note that as early as 1923, Rudolf Steiner indicated that if beekeepers continued this breeding practice, then the improvements they had seen would barely last a hundred years (*Bees*, p. 178). In view of the bees' present-day situation, this has already proved true.

It is certainly true that breeding has brought about significant improvements – we would not wish to overlook this. But neither should we ignore the downside of breeding. This downside probably lies in the fact that breeders have simply neglected the essential nature of the bee. Beekeepers were intoxicated by their success: colonies had to become larger and the honey harvest had to increase. And of course, the bees were no longer supposed to sting. The swarming instinct, the natural way of reproduction and rejuvenation,

The goal all beekeepers should aim for is to help bees live naturally.

was reduced to a negative phenomenon. Natural queen rearing by colonies was suddenly undesirable. Human beings started to fumble in the handiwork of creation. We can find the result of this today in the many diseases and weaknesses of our bees.

The goal all beekeepers should aim for is to help bees live naturally. It is true that in order to quickly spread a standardised bee through a region, it can only be done via breeding, at least initially. However, thereafter it would be necessary to give natural reproduction and rejuvenation the importance it once had for bee colonies.

These issues should not be regarded as the only cause of the present situation in beekeeping, but rather as a part of what arguably would have to be changed in future.

Many beekeepers find queen breeding indispensable for various reasons, and, if they rely solely on natural reproduction, are not able to obtain the many queens they need to increase their number of colonies quickly enough. There are many ways to raise a large number of queens, but they all rely on two basic methods: breeding in queenless colonies and breeding in queen-right colonies. Those only wanting to learn the practice of breeding should start with breeding in queenless colonies.

Breeding in queenless colonies

Queen breeding, or more correctly, queen reproduction, requires a *breeder* or *breeding colony* and a *queen rearing* or *raising colony*. The breeder colony is the colony out of which the breeding material is taken – that is, the small worker larvae. The rearing colony is the one in which the queens are reared or raised from worker larvae.

The breeder colony should meet certain conditions before you can breed from it. The colony should already have been observed for two years and all observations recorded in writing. It should be hard-working, gentle, robust and stay on the combs, as well as exhibiting harmonious growth. By staying on the combs, we mean that when the hive is opened, the bees remain quietly on the combs, even when the combs are removed from the colony. The colony should not be susceptible to disease or sensitive to changes in the weather. Furthermore, it should be true to the race and fulfil the requirements of the selection guidelines for breeding purposes.

The rearing colony should be a colony of above average strength with a large and very good brood nest. It need not fulfil any breeding requirements as it is intended only to feed and care for the larvae. The rearing colony needs lots of young bees and nurse bees. As we do not know whether the feed also imparts behavioural characteristics, colonies with extremely pronounced characteristics, such as an inclination to sting, swarminess or disease susceptibility, should not be used for rearing.

As queen breeding does not allow us to make use of a natural queen cup, we have to make them ourselves out of wax. The plastic cups available from suppliers should be avoided. Wax cups are made with a cell-moulding tool. This is a round, wooden rod, rounded at one end, with a thickness of about 9 mm (⅜ in) and a length of 10 cm (4 in). It should be completely smooth so that the wax can be easily removed from it. Warm the wax in a bain-marie until it just melts. Dip the moulding tool in cold water, remove it and shake off any drops of water. Now dip the moulding tool into the wax to a depth of about 10 mm (⅜ in) and remove it. Repeat this three or four times, each time dipping the tool a little less deeply into the wax. Finally hold the tool with its wax cap in cold water and pull off the completed queen cup. If, despite the wetting, the cap is difficult to remove from the moulding tool, some honey may be added to the water as a lubricant. Lubricants used in the production of foundation should be avoided so as to save the bees unnecessary cleaning work. It is more pleasant for the bees to lick off residual honey instead of lubricant.

When sufficient cups are moulded, they need to be left to dry properly. Next, they are pressed into the hollows in cell plugs so that they make the best possible connection with them. You just need to be a little careful with the cups so that your handling does not deform them. The cell plugs with their cups are now pressed into a cell holder with the cups facing downwards and mounted in the breeding frame. These preparations can be done any time before grafting the larvae.

Preparing the rearing colony starts nine days before grafting the larvae. A rearing colony is a strong colony with a good brood nest. As we are rearing in a queenless colony, we need capped brood as well as many nurse bees. This can be achieved by hanging the frame on which the queen has been found above the queen excluder in the honey chamber. If there are no more empty combs in the honey chamber, but instead combs largely filled with honey, up to two of these honey-filled combs are exchanged for empty, light-brown combs moistened with honey. These light-brown, honey-moistened combs will be quickly cleaned by the worker bees and the queen can then begin to lay up a new brood nest in the honey chamber. Within the next nine days, a wonderful, new brood nest is created. This new brood nest ensures that there is now a large excess of nurse bees, because in the lower brood chamber the old brood nest is still present, but its larvae were all capped within the nine-day period. Once the nine days have passed and no open larvae can be found, queen breeding may begin.

The queen in the honey chamber is caught and caged. She can later be used for nuc making or be given to another colony. All combs are removed from the honey chamber and placed on the comb rack. The honey chamber is now removed, or, if using a rear-access hive, closed up so that bees cannot enter the empty space. Now brush into the lower brood chamber the bees on the combs that were removed from the honey chamber. These combs primarily contain nurse bees needed for raising the queens. Thus, in the brood chamber we now have capped brood combs and lots of young bees to look after the brood. The brood chamber combs are once more checked for any queen cells that may be present. If there are any they must be removed, otherwise the queen reproduction in this colony will fail.

All these preparations should be carried out when both Venus and the moon are in a light/flower constellation (Gemini, Libra or Aquarius).

To increase breeding success, and above all to be able to graft larvae into clean cups, the breeding frames with moulded queen cups are placed in the rearing colony in the location where they will later be hung for rearing. The bees can now clean the cups perfectly before larvae are grafted into them. There should be brood frames on both sides of the breeding frames so that an even temperature is maintained. Before grafting larvae, the breeding frames should of course be removed again.

The breeding colony also requires preparation. On the sixth, fifth and fourth days before grafting, its brood nest is given a light-brown comb moistened with honey so that on the grafting day there will be enough young larvae available. These three combs are given on the sixth, fifth and fourth days because it may happen that the colony does not immediately accept the combs, but allows between one and one and a half days elapse before doing so. The larvae to be used for grafting should be one to two days old at the most. That means that the queen has to lay on the combs four to four and a half days before the grafting date.

Grafting is generally done with a grafting tool. It is also possible for this purpose to use a pointed piece of wood or a very small watercolour paintbrush, or even an appropriately trimmed quill. If you settle for the grafting tool, the Swiss grafting tool is especially convenient. It is angled at the end to make it easier to lower it into the cells and lift out the larvae. Grafting tools produced by machine are frequently so thick at the tip that they are difficult to insert under a small larva and you just turn the larva round rather than getting it onto the tool. For this reason, the tip should be rubbed thinner with very fine sandpaper and the angles rounded off. If you have never picked up larvae, it is advisable to practise on a dark comb. Light combs are still very soft. When you try to lift out the larva with the grafting tool, it is easy to pierce the cell floor, thereby injuring the larva in the cell on the opposite side. It will not survive this injury.

It is preferable to graft with the youngest larvae. Older larvae have been in a hexagonal cell for longer and have therefore absorbed more of its forces. Furthermore, with increasing larval age the ovarioles of the resulting queen will be fewer in number and this will reduce her laying capability.

Once you have grafted larvae into all the prepared queen cups, put the cell holder back in the breeding frame. About twenty queen cups can be fitted into a breeding frame, but you should not expect a queen rearing colony to look after more than two breeding frames.

If at the time of queen breeding there is no honey foraging to enhance the willingness of the colony to tend the cells, 2–5 litres (2–5 quarts) of honey syrup may be given in a feed bucket or dish. This should be given in the evening of the day on which the larvae were grafted, because otherwise the expected success may fail to materialise. Larvae that are not nursed by the bees from the first day onwards will also not be nursed on the following day merely by giving feed. They will either be eaten or thrown out. The willingness to care for the larvae is also influenced by the available stores. Only a colony with good stores nurses well. A helpful way to judge this is to say that 20 per cent of the combs in the rearing colony should be full of stores, not including the arc of stores over the brood. As nurse bees need a lot of pollen for their brood-food glands to function properly, we also need to ensure that there is enough pollen in the rearing colony. Excess feeding, however, does not mean that successful nursing will greatly increase. Precisely the opposite can happen if too much feed is supplied to the colony.

Based on these observations, we can say that the willingness of the nurse bees to tend the larvae, and therefore the success of queen reproduction, depends on guaranteeing that there are enough young bees, that the work is carried out under the right constellations, and that there is the necessary flow of feed and availability of stored pollen.

Three days later, the developed larvae may be counted after carefully removing the breeding frame. We do not need to look into the queen cells, but can see from the extensions of their cell walls whether the larvae have been accepted. By this time, the bees will have made a pitcher out of the cup by lengthening its wall. The breeding frames are replaced, and the rearing colony closed up. The colony with the cells is left completely alone until shortly before they are due to hatch. When we know the number of queen cells available, we need to consider what we want to do with them. Breeders let them hatch in incubators, but as most beekeepers don't have an incubator, they let the queens hatch from their cells in cages. To do this the cells ready to hatch are put in a large cage with about ten to fifteen bees and given some feed. The cage is then placed in the honey chamber above the brood nest on a mesh through which no bees can penetrate. So that the cells will benefit from the warmth of the brood nest, the remaining empty space in the honey chamber is filled with rugs. Between one and two days later, the queens should have hatched. The hatched queens are now transferred to so-called small mating colonies and, after two days in darkness, moved to the mating apiary. This is the where there are several parent colonies which, due to their increased supply of drones, guarantee the best possible mating.

As regards the hatching queen and the bees for whom the queen is intended, it is certainly best if the queen is allowed to hatch in the colony where she later will be active as its mother. If the queen breeding is done in order to make up nucs, it is best to use the cells for making the nucs without first letting them hatch.

To obtain cells that are really ready to hatch, a small precise calculation needs to be made. A queen needs about sixteen days from egg laying to hatching. The larvae that were grafted into the queen cups were one day old. We must allow three days at the egg stage, which means that the grafted larva is four days old. The right time for processing the cell would be on the fifteenth day of the entire development of the queen. The fifteenth day is thus the eleventh day *after* grafting. As it can happen that a grafted larva is already somewhat older than one day, to be on the safe side the cell processing should occur on the morning of the eleventh day after grafting. If we miscalculate by one day – something that can happen to any breeder – it may turn out that the first queen to hatch in the rearing colony (as already described in the previous chapter) starts tooting and finds one hatch-ready cell after another and stings their occupants. Our efforts with propagating queens would then have been in vain.

However, we should not be over-anxious and process the cells one or two days earlier, on the thirteenth or fourteenth day of the queen's development. It is precisely at this time that the cells are sensitive to changes in temperature. If the cells cool down, queens will still emerge from them on the sixteenth day, but often their wings will not be fully developed. This makes it impossible for them to go on their mating flights. Successful queen reproduction therefore depends on adhering closely to the rules given above.

To make nucs with hatch-ready queen cells, it is advisable to use two queen cells for each nuc. This is because one of the two might not hatch. Nuc making then proceeds as described in the previous chapter in the section on making a nuc with capped brood and a single queen (see pp. 80). In the present case, instead of an already mated queen, we provide two queen cells. They are either embedded in a brood comb or fixed with matchsticks or wire to the comb. We must avoid pressing on the cell during these manipulations. The cells are embedded or secured in such a way that the bees can keep them at the right temperature and the queen has enough room below the cell to hatch out. If the queen cell is secured to the comb with crossed matchsticks or a piece of bent wire, the comb's cells behind the queen cell should be pushed in somewhat so the bees can get around the queen cell to keep her warm.

Processing queen cells should take place very quickly, therefore the nuc is brought as far along in readiness as possible so that our last task is giving it the queen cells. The nuc is then left alone for about three weeks. Thereafter, if the weather has been favourable, the nuc should have brood in it.

The nice thing about this way of making nucs is that the emerging queens are guaranteed to be accepted by the bees, and from the moment of hatching they ensure harmony in the nuc.

As the queen has still to be mated, it is advisable not to position the nuc among a great frontage of hives, but rather to site it alone at the side. This will largely avoid queen drifting.

Breeding in queen-right colonies

The rearing of queens in queen-right colonies is another option for queen reproduction. As criticism of artificial queen breeding has not yet ceased, some beekeepers have tried to make it more like natural queen reproduction. Queen reproduction in queenless colonies is of course very similar to the development of emergency queens. As it is generally known that supersedure queens can be very good queens, people have tried artificially to reproduce the supersedure mood. It is done in the following way. First, it requires a colony in a magazine hive with plenty of brood. The queen is confined to the lower box with a queen excluder. The brood nest is moved to the topmost box, the third box up. The queen excluder is covered with foil so as to leave open a narrow strip of 3–5 cm (1–2 in) as a passage through to the upper boxes. This greatly separates the queen and her brood nest, and by covering most of the queen excluder there remains only a little contact between the lower and upper boxes. The breeding frames with larvae are now hung between the combs in the brood nest that is now in the top box. The bees in the top box are so far from the queen that they actually accept some of the larvae and rear them into queens. But in terms of numbers, queen-right colonies raise fewer queens. However, with this method of queen rearing, the upper brood nest with its breeding frames needs checking for emergency queen cells after four or five days. If the connection with the queen in the lower box is too small, it may happen that the brood nest bees assume they are queenless and, to be on the safe side, make emergency queens. These have to be cut out as they will emerge sooner than the queens from the breeding frame, thus again risking the success of rearing, because the first emergency queen to emerge will bite into the other mature cells and sting the queens inside them.

Beginners in queen breeding will find it easier to learn breeding in queenless colonies and thereafter in queen-right colonies. Whichever method you finally adopt rather depends on whether you need a large number of queens, or whether a few will suffice.

These two accounts of queen reproduction in both queenless and queen-right colonies is only intended to show how it is possible to rear many queens. We have not discussed the various ways of bringing about better care of the developing cells. There is already sufficient published material that deals excellently with this theme.

Queen reproduction through deliberate use of the swarming instinct

In order to reproduce queens without using worker larvae, there is the possibility of rearing queens from eggs laid in round cells. This is done by deliberately using the swarming instinct. For this we choose a colony that is worth breeding from and we wait for the swarming time to occur in the apiary. During this time, when colonies are normally given space to grow in order to subdue the swarming mood, we let the chosen colonies remain cramped. At the same time, we feed them with flower honey if there is no nectar available. Deliberate stimulation of swarming works best with rape or buckwheat honey.

If a colony is in a swarm mood we need to establish when the first queen cell is ready for capping. On the day it is capped, we remove from the colony the old queen with a brood frame and its bees, and two honey frames and its bees. With the old queen we start a so-called queen nuc, because we want to keep this queen as valuable breeding material. The further procedure has already been described in Chapter 6 in the section on artificial colony increase (p. 70)

This kind of queen reproduction is more satisfying, not only for the beekeeper but also for the bees, who experience the development of the queen right from the start. From the queens hatching in the small nucs, we will get fine, vigorous and harmonious colonies that by autumn will certainly occupy a whole hive box. We may wish to make one or two more nucs than we actually need, because there is always the risk that a queen drifts or gets lost somewhere on her mating flight. If one queen fails, the queenless nuc is then united with a queen-right nuc.

Those who have not yet tried this form of queen reproduction will assume that it is significantly more onerous than artificial queen reproduction. In my experience quite the opposite is the case. Once you have learned to recognise the various stages of queen larvae, the expenditure of effort in queen reproduction is significantly reduced. In order to learn to recognise the different larval ages, you have to make very careful observations. It is easiest when a queen cup is being built in the building frame. Then you can watch it each day as it develops into a capped cell. The colony will certainly not be pleased if you open the hive daily for nine or ten days. However, if you have trained your eye, you will be able to do a lot for the bees with this more bee-appropriate, bee-friendly form of queen reproduction.

Chapter 8
Honeycomb Construction

The construction of combs demonstrates the bees' skill and efficiency as builders. They show us how they can create a strong yet light structure from a relatively soft material like wax, and then how they can shape and arrange the cells in such a precise and orderly way that they take up the least amount of space.

It is generally said that bees build these hexagonal cells in order to make the most of the space at their disposal. This is true. But from Rudolf Steiner we know that this is not the only reason. He explained that certain forces have been given to bees to enable them to build such precise cells, and that these forces are the same ones present in the earth that create quartz crystals (*Bees*, p. 51). Quartz crystals are also hexagonal. Steiner further explained that the bees must be raised in these cells if they are to be capable of building them, as the structure of the cell imparts these forces to the developing bee. He said that even in the earliest stages of development, during the egg and larval stages, the bees receive from the cell the most intense effects of these forces (*Bees*, p. 7).

The substance used to build these cells is wax. The wax in this case is an endogenous substance, meaning it is created within the bees themselves. On the underside of a bee's abdomen are glands that, in young bees, are developed to produce wax. This wax can be found and examined in the form of small, white scales, that accumulates, for example, in the winter debris on the hive floor. Thus, no exogenous, or external, substances are used in cell and comb building. In older bees the wax glands have already ceased their function, but if they are put in a situation where they have to build again – for example, in the case of artificial swarming – then these wax glands can resume their functioning.

There are three different cell types within the bee colony in keeping with the three different castes of bee, namely queen, worker and drone. Queen cells are round in section and look very similar to a grape. Worker and drone cells are hexagonal in cross-section and, in their capped stage, look like a quartz crystal with points at each end.

Queen cells occur in isolation in colonies. They are built in the form of swarm cells and supersedure cells at the comb margins; they can also be found as emergency queen cells in the middle of brood combs. Worker and drone cells are usually assembled into sheets of comb, meaning the cells are built next to each other, thus forming the surface of the comb. The combs are built vertically and the cells are arranged on either side, front and back. Normally worker cells and drone cells are confined to separate combs, but at swarm time the bees tend to add drone comb to any free space adjoining worker comb. Worker cells and drone cells have the same shape, they are distinguishable only by their different cell sizes. As a worker is significantly smaller than a drone, her cell is also smaller. This difference is so great that anyone can see it at first glance.

In individual colonies there are different areas of drone comb. If colonies are given the freedom to build comb, it can happen that one colony has only one drone comb whereas another builds three. We cannot tell just by looking how much drone comb an individual colony needs. The colonies focus wholly on their own requirements, and therefore make very different amounts of drone comb. If the beekeeper wants to control or influence the amount of drone comb, they should in any case allow at least 10 per cent of the entire comb to be built as drone comb.

Natural comb construction

Natural comb construction occurs everywhere that beekeepers let their bees build comb without frames or foundation. This was part of traditional beekeeping (also known as Zeidler beekeeping, where colonies were housed in the cavities of trees) and of the many different forms of skep or basket beekeeping. Only when the frame of August von Berlepsch (1815–77) was invented in 1852, and Jean Mehring (1816–78) developed his foundation, did beekeepers start to turn away from natural comb. At the same time, the development of foundation and frames changed the habitations of bees. Skeps declined in use and wooden boxes began to take over in beekeeping. Nowadays natural comb occurs only as drone comb in colonies, but we can still find and admire natural worker comb in small mating nuc colonies and in the relics of basket beekeeping, such as is found on the Lüneburg Heath in Lower Saxony, Germany.

The ability of bees to build cells and construct combs is very important to them. In his lectures on bees, Rudolf Steiner made a very interesting comparison between the human being and a bee colony. He compared the queen bee to albuminous cells, worker bees to the blood and drones to the nerves, and he likened the human body to the bees' comb. Thus, Steiner indicated that the comb is the colony's body (*Bees*, pp. 16–18). With this in mind we can see how important it is to the bees to be able to make their own comb. In doing so, they are creating the colony's body using wax that they make in their own bodily organism. During an early stage in their development each bee spends part of her time as a comb-building bee. This means that each bee participates in the building and maintenance of the colony's body through the capacities of her own body.

During an early stage in their development each bee spends part of her time as a comb-building bee. This means that each bee participates in the building and maintenance of the colony's body through the capacities of her own body.

The description of the comb as a body is particularly interesting to those who know something of beekeeping jargon. When beekeepers refer to a bee colony, they almost always have in mind a colony with combs, but a colony *without* comb is referred to as a naked colony or a naked swarm (although the term is not reserved exclusively for when the colony is not in the hive). The term naked colony relates to the lack of corporeality of the entire colony, as found in both artificial and natural swarms.

If we look at the human body, we see its typical human form. If we examine a bee colony's body, its comb assembly, we notice that it is quadrangular. Should we therefore assume that along with the round shape of the queen's cell and the hexagonal shape of the worker and drone cells that the bee colony has a third shape, the quadrangular shape of the body of the colony itself?

Not necessarily. To discover the real shape of the colony's body, we must allow a swarm to build as it wishes. To do this we tip it into an empty hive, provide it with feed as described in Chapter 4, and let it build comb without frames or foundation. After one or two weeks we can see the results. Certainly, most observers would be surprised by how different the shape of the colony is from what they had previously assumed it to be.

The swarm does not want to build totally straight comb surfaces – natural comb can be quite wavy or angled – but neither does it allow itself to be wholly influenced by the shape of the hive. Instead it builds its body in a rounded shape. This is regarded as perfectly normal in skeps because the swarm cannot build otherwise, but you may be surprised that even in a square or rectangular box a swarm prefers to make a round shape. However,

FIGURE 33. This swarm did not find a habitation and built its entire comb in the open. The rounded shape is clearly visible.

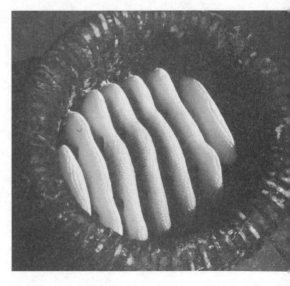

FIGURE 34. Natural comb in a Lüneburg skep.

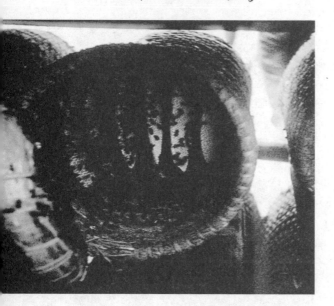

FIGURE 35. Natural comb in a Thüringian roller – a horizontal, cylindrical hive. This is a rare picture as these basketwork hives can usually only be found in museums.

if you give a swarm some foundation for building, this round shape becomes less clear, detectable only with difficulty. The fact that the round shape means a lot to the bees (Rudolf Steiner called it a sun-like shape) is evident in the construction not only of comb or queen cells, but also in the brood nest itself, which always has a rounded shape. This shape can be observed not only vertically on the individual combs, but horizontally across the whole brood nest. There can be no talk therefore of a third kind of shape in the bee colony, namely the quadrangular one. This shape is imposed on the bees and is not typical for them. Regarding the form of the human being, we have a body with a head and four limbs. But we can also see in the human body a five-pointed star: feet apart, arms outstretched, head upright. Turning now to the bee colony, we see a body that comprises organically arranged combs. These are so constructed and arranged that we can see in them the shape of a sphere. The bee colony always tries to obtain a sun-like, round or spherical corporeality. This is hindered simply by the shape of the hive. Even so, we can still find in the brood nest the tendency towards roundness, which can be seen in the oval shape if the frame or foundation is not square but rectangular. This deviation from the round shape to the oval shape can also be seen in skeps if the height or length is greater than the diameter. The German beekeeper Ferdinand Gerstung (1860–1925) observed colonies on various comb sizes and discovered that the brood nest of conventional bees has a diameter of about 26 cm (10 in). This size should be strictly observed and never reduced if you tend hives and want to take into consideration such subtleties.

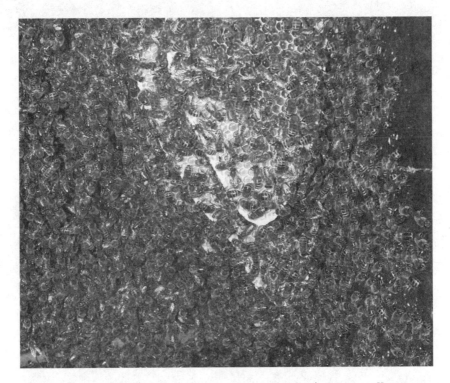

FIGURE 36. *Natural comb (that is, foundationless) in a small cast – a secondary swarm or after-swarm. The comb is off-centre, more to the left side of the hive cavity. The environmental warmth affects where a colony starts building. In this case there was another hive against the left-hand side of the hive. This would have helped warm that side of the hive and the bees obviously chose to build nearer that source of warmth.*

 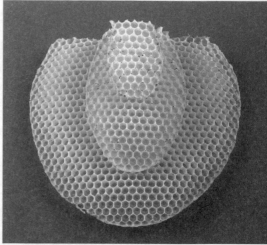

FIGURE 37. This swarm has not succeeded in making one big comb out of two heart-shaped combs. In the middle the cells do not match up. I would not breed from this colony.

FIGURE 38. Heart-shaped combs positioned one in front of another, just as intended by a swarm. The 'dirt' on the cell margins is propolis.

FIGURE 39. Frames with worker comb (left) and drone comb (right), each with some hatched brood.

Foundation

We owe foundation used in hives to the keen observation and skilled craftsmanship of Jean Mehring, a master joiner. He observed that when bees build comb, they do not make completed cells one after another. First, they make a small wax sheet, then they make a series of hollows in it that later on become the floors of the cells. Finally, they draw out the cell walls round these hollows.

These observations led Mehring to the idea of providing bees with wax sheets that already had the cell floor hollows formed in them, so the bees only had to draw out the walls. He constructed a prototype of today's foundation presses and used it to make wax sheets. When he placed the sheets in his hives, he found that his ideas worked. Since then, foundation has increasingly been used, and today's beekeeping is unimaginable without it.

The use of natural comb and comb built with foundation

Considering the section on natural comb (p. 105), the question might arise as to which method of comb building is the more bee-friendly. In order to find an answer to this, we need to look more closely at natural comb and foundation. We have discussed the round shape of both the colony's body and of its queen cells. We have also mentioned how much the bees value having queens that are able to develop in round cells from their first day, unlike emergency queens that are raised from converted worker cells that retain something of their original hexagonal shape. Indeed, Rudolf Steiner implied in his lectures on bees that this round shape is a sun-like shape (*Bees*, pp. 7f). If we recall what was said earlier about how a particular shape can exert a particular influence, we can see that the round shape of the cell exerts its round or sun-like forces on the growing queen. A queen who, like an emergency queen, has not been able to take in these sun-like forces over the entire period of her development, seems to the bee colony not to be a queen in the fullest sense.

Just as the round cells exert forces on the queen, so also do the hexagonal worker cells exert a totally different effect on the workers. Rudolf Steiner indicated that the worker can build hexagonal cells with such precision only because she has grown up in such a cell and absorbed the effects of its forces (*Bees*, p. 7). These formative forces may also be described as hexagonal forces.

Changing over to natural comb

Those who want to enable their bees to make natural comb will soon realise that it is not easy to *re-accustom* colonies to natural comb construction in the form of worker comb. If you let the bees build naturally instead of giving them combs or foundation, they lapse into a veritable euphoria of drone-comb building. They make incredible amounts of drone comb that they do not need.

This drone comb issue is similar to the situation in agriculture when a farmer wants to switch from conventional farming to organic or biodynamic farming. By stopping the use of herbicides, the farmer may notice that the soil produces enormous amounts of herbaceous plants in the first couple of years, which, due to their numbers, become weeds. The soil seems to want to facilitate plant growth that previously was suppressed by herbicide use. The bees appear to behave in a similar fashion when they can suddenly build as they wish, producing excessive amounts of drone comb. However, with proper colony management the amount diminishes.

In order to prepare colonies for natural comb, it helps to leave them with one or two drone combs. As soon as swarming time is over, the colonies can gradually be changed over to natural comb. To achieve this, it is advisable to remove the queen excluder and fill two thirds of the honey chamber with frames that contain only half a piece of foundation cut diagonally. As these half sheets of foundation are not as secure as whole sheets, they are soldered on both sides to the top bar with melted wax in the same way that starter-strips are soldered on. The remaining third of the honey chamber is filled with straw dividers. Only two-thirds of the honey chamber is used, so that the frames will be fully built up. Later the rest of the space can be released for building.

The hive in which natural comb is being built should be exactly level, so that the bees embed the wires of the frame within the central walls of the comb. The frame wire makes the resulting combs so secure that they may be centrifuged. If the transition from foundation comb to natural comb has been practised for one or two years, you may give starter-strips instead of half sheets of foundation, as described in the section on swarming, and thereby acquire totally natural combs.

Natural comb starting from starter-strips is often confined only to the brood chamber. It is mostly built in such a way that it cannot be centrifuged. This is because it is insufficiently built out to the side and bottom bars. In order to obtain natural combs that can be centrifuged, we must proceed in the following way. We begin with a well-built-out, light-brown comb. This is cut out so as to leave about three rows of cells on the top bar, and half a row of cells on each of the side and bottom bars. The remaining cells are now bevelled to the septum (the cell floor) with a knife so the bees have a good start for building. Provided that the hive is level, the bees will make the best natural comb right round the remaining cells. In the space above the brood nest – far from the entrance – you will get the best natural comb. The combs to be used for cutting out should be clean and, to avoid spreading disease, not from colonies suffering from nosema or dysentery (evidenced by faecal splashes).

If new frames with starter-strips are used, the bars must be coated with wax on the inside surfaces in order to obtain the largest possible area of cell building in the frame.

If the bees suddenly change from worker comb to drone comb, the comb building can still be directed by cutting away the drone comb. The last row of worker comb should also be removed with it, leaving not a single drone cell. If even only a small part of a drone cell remains, drone comb construction will be resumed from this point. The edge of the cut is then bevelled. After this, the bees will continue with building worker comb. This normally only works, however, before and after swarming time. A colony that is in swarming mood will increasingly build drone comb, or stop building altogether.

The queen's age also plays an important part in the transition from comb built on foundation to natural comb. Clear differences can be observed between the quality of construction by a prime swarm with an old queen, and that by a cast, or secondary swarm, with a young queen. Casts generally produce better natural comb than prime swarms. It is therefore advisable to start with colonies that have a young queen because these colonies are less prone to enter swarming mood and more inclined to produce natural worker combs.

Building in the honey chamber

We might still ask why do colonies build so well in the honey chamber and why is the queen excluder removed from the colony? As has been mentioned, the colony aims for a round bodily shape. If the frames to be built up are placed in the honey chamber in the middle of the hive, the colony fills the frame from the top bar to the lower brood chamber combs and the bottom bar in order to achieve a closed comb surface. The same effect is less easily obtainable in the brood chamber, because a large area free of comb is not felt by the bees to be as disturbing as it would be in the middle of the existing combs. Because the queen excluder represents a significant foreign body in this assemblage of combs, it should be removed during the building process to facilitate the 'growing together' of the comb.

We can promote the bees' inclination to build by using the cosmic effect of earth/root times, and on these days carry out the processing resulting from the natural comb building. The procedures thereafter should be reconnected to light/flower times or warmth/fruit times. If the colonies are successfully transferred to natural comb, we will subsequently have to provide our own bee material for queen rejuvenation. If we constantly buy in new queens we will find it very difficult to maintain the urge to build natural comb, because the progeny of the bought-in queens will first have to be introduced to natural comb building. Therefore, to attain a good bee, I believe it is worthwhile doing the selection work for breeding in our own apiary.

The age of foundation wax

These days we take replacing combs with foundation for granted. With the help of foundation, it is possible to obtain slightly more comb than can be expected with natural comb building. Except for in the middle of the brood nest, foundation can be put anywhere in the colony, and, when inserted at the right moment, will be built up. A big advantage is that bees build mostly worker combs on combs built from foundation and almost no drone comb at all.

To better understand the change from natural comb to comb built with foundation, we need only look at the production procedure from old combs to foundation. Beekeepers either make their own foundation, or they give their old combs to a wax rendering company and receive in exchange manufactured sheets of foundation.

The old combs are melted in a water bath or with steam. The resulting soup, consisting of wax, cocoons and faeces, is pressed to remove the liquid wax from the solid substances. The wax is purified several times until all particles of dirt are removed. Foundation sheets are then pressed or rolled out from this wax.

The wax used for making the foundation has to be heated for half an hour to about 120°C (250°F). This procedure is necessary for killing the particularly resistant spores of foul brood (this is discussed in more detail in Chapter 11). As old combs are repeatedly melted down and processed into foundation, it easily happens that a part of the wax of a sheet of foundation has been through this sterilisation process many times, and therefore in sheets of foundation we end up with not only new wax, but also some very old wax.

We know that wax is a very tough substance, but do we know what subtle changes take place in the substances of the wax when it is heated to such a high temperature? Is foundation wax of the same chemical quality as wax coming fresh from the bees? It is hard to imagine so. When considering the creation of the colony's body, the question must arise as to whether such a body that is partly made from 'overheated' wax is capable of offering the bees the qualities that natural comb makes available.

By comparing natural comb with foundation comb, it can be seen that the former has a significantly thinner septum, or wall, that divides either side of the comb (in German foundation comb is known as *Mittelwand*, or 'middle wall'). This results from the fact that foundation has to be made relatively thick to withstand the various procedures up to and including fixing it into a frame. Because the bees do not want such a thick septum, they gnaw away at the excess wax and use it to build the cell walls. This means that the cell walls of a comb drawn on foundation are not made entirely of newly produced beeswax. Instead, going from the foundation outwards, about one third of the wall area comprises old wax, and only the outer two thirds of the cell walls are made of fresh beeswax. The difference between the old foundation wax and the freshly produced beeswax is clearly visible in its coloration. Foundation wax is generally yellow, whereas freshly secreted beeswax is almost perfectly white.

Rudolf Steiner indicated that bees take up the hexagonal forces of their cell envelope most intensely during the egg and round larval stages of their development (*Bees*, p. 7). If we consider that substances also affect living organisms, we are justified in asking whether bee development is influenced differently by comb made from foundation than from natural comb. In view of the fact that bees are developing to the round larva stage in a substance that is basically made of old wax that was heated to high temperatures, it is easy to imagine that comb made from foundation does not exhibit the same properties as substances found in natural comb.

This means that giving bees foundation leads them not only to reuse old material for building the colony body, but at the same time to build cells whose substances cannot transmit the same effect as those found in natural comb, even though the cells are the same shape in both. Those who would help the bees build a healthy colony body should return to natural comb; those whose focus is on simple and uncomplicated beekeeping will settle for foundation.

Conclusion

The foregoing discussion of natural comb compared with comb made from foundation makes it clear that, for the bees, there is a considerable difference between these two methods of comb construction. Those who recognise and accept this should keep in mind the challenges of going from comb based on foundation to natural comb. Returning successfully to good natural comb demands not only patience from the beekeeper, but also a capacity for empathy with the bees. Anyone who takes this course may feel that they are actively working for the health of their bees.

It has already been indicated that bees build well in the honey chamber. In connection with this I would like to mention something that really ought to be dealt with in the chapter on honey, but as it has more to do with freshly built comb, I mention it here. Some beekeepers think that they should always aim to use young combs in the brood chamber, and in the honey chamber use combs that they no longer expect the colony to put brood in. But when comparing the taste of different honeys, my belief is that the combs in the honey chamber should be just as young as those in the brood chamber. Honey stored and ripened in white or light-brown combs tastes better than honey that ripens in brown or even black combs. If you want to harvest honey of perfect taste, it is only possible from light-coloured combs. The deterioration in taste in dark, old combs arises not only from the dirt particles that adhere to the combs, but also from the cocoons and the larval faeces deposited under them.

The colour of a comb depends not only on its age but also on how many brood cycles it has been through. Freshly built comb is referred to as white comb. It is very light coloured and has not yet had any brood in it. Combs that have had a few brood cycles in them become light-brown and, after many such cycles, turn from brown to black. These combs should be removed from the colonies and rendered. Each bee that hatches from a cell leaves behind a cocoon, thus somewhat narrowing the cell. This means that smaller bees emerge from cells that have had many brood cycles, compared with cells that have seen few brood cycles.

In modern beekeeping, where the beekeeper can influence the age and condition of the combs, they will swap out the dark combs. In natural comb, for example in hollow trees, the bees do this themselves. When they sense that the cells have become too small, they scrape away the cell walls back to the septum, or cell floors, and rebuild new ones. Observations have shown that this takes place every four or five years, and it is interesting to note that in colonies that are left alone, this four to five year rhythm coincides with the lifespan of a queen. Comb that needs to be renewed usually only concerns worker comb cells in the brood nest, because only brood nest cells shrink through having had brood in them. Here we can see a good connection between the queen and the brood nest because it is precisely here that the queen undertakes her most important task. The whole brood nest is the realm of the queen, not only because of the brood itself, but also because we find in it very clearly the sun-like roundness in which the queen, a creature of the sun, lives and devotes herself to completely.

Beekeepers working with modern races or racial crosses will not agree when we treat as equivalent the queen's age and the age of the comb. Those who keep Carniolan bees (a subspecies of the western honeybee native to Slovenia and surrounding countries) will know that the tendency to swarm is very strong, and that a queen rarely remains in a colony for four or five years before leaving in a swarm. Furthermore, hardly any queens nowadays live for four or five years. If they do, they would be good breeding material that anyone would want to keep at all costs. However, as these observations with a four to five yearly rhythm stem from a time when we still had native bees, we need to consider the behaviour of such bees. As long as they were not hybridised, the behaviour of the old races was such that their tendency to swarm was much less pronounced than is found in some bees today. These old, native races were more strongly inclined to supersede their queens than is seen in our modern bees. Thus, the parallels between comb building and the lifespan of a queen is thoroughly appropriate.

Chapter 9
Honey

After the bees swarm there usually follows the time of the honey harvest, although there are regions where the honey harvest starts before or during swarm time. In order to harvest the honey it is important to know where the bees are storing it. Here we notice a fundamental rule from which bees rarely deviate, which is that the honey is stored as far away from the entrance to the hive as possible. If the entrance of the hive is at the bottom, then the honey chamber will be at the top, with the brood nest below it. Bees have a certain order for the placing of their stores.

Apart from the brood and honey, pollen is also stored in worker cells. If we look first at a brood comb, we can see the roundness of the brood nest and above that, further from the entrance, an arch of pollen – assuming once again that the entrance is at the bottom of the hive. Close to the pollen, and further still from the entrance, there is an arch of feed or honey. This arrangement shows that the comb used in the brood chamber should, if possible, be big enough to accommodate not only the brood, but also the pollen and some honey. The honey that is stored outside the brood nest is stored in the *honey chamber*. To prevent the queen laying in the honey chamber, she can be separated from the brood chamber with a queen excluder. Only worker bees can pass through the excluder.

There are some regions that have only very small or very special early nectar flows that beekeepers wish to make use of. Sometimes, however, the honey chamber is too big and the colony not yet strong enough to occupy it. The small amount of honey then becomes distributed over many combs and cannot be extracted by centrifugation. In order to take advantage of the smallest flows, one or two light-brown empty combs may be hung behind the inserted cut-out building frame, on the side furthest from the entrance. The bees will then carry the honey into these combs, which can be removed and centrifuged when the honey is ripe.

Trough hives, sometimes called long deep hives, are particularly suited for this procedure, because, as the name suggests, they can be extended rearwards as required. Top access hives, such as magazine hives whose space is limited, have half-depth honey supers, particularly for early flows and for small, special flows. Two of these supers, one above the other, make up the same height as occupied by whole combs. In half-depth supers, one may use half-height combs or thick combs.

In the previous chapter, the different types of cells found in a bee colony were described: the round queen cells that occur alone depending on the time of year; the hexagonal worker cells, assembled side by side across whole surfaces, that make up the overall comb structure; and the hexagonal drone cells that are likewise put together in large numbers to form parts or whole surfaces of comb. Neither honey nor pollen are stored in queen cells. As soon as the queen has hatched, they are generally dismantled. Worker comb is where the honey and pollen are stored and ripened.

Drone combs are not usually used as storage for honey and pollen, except when there is a surplus supply of nectar and there are no more empty worker cells. Then they are used as a sort of interim store. If the honey in the drone comb is not capped, then either the honey is used by the bees or it is moved to another storage site. This indicates that bees clearly prefer worker cells for storing honey and pollen, whereas the use of drone cells to store pollen has yet to be observed.

This preference can be clearly seen on the building frame. If a colony is given a building frame and is not yet in a swarming mood, then it may begin to build worker comb. If a period of bad weather follows, then construction temporarily stops. It is resumed again when the weather improves. During the period of bad weather, however, the urge to swarm begins to build and as a result the colony no longer builds worker comb, but switches instead to building drone comb. We therefore see mixed comb in the building frame. If a honey flow now starts, we may notice that nectar is also stored in the building frame, as long as its comb is not already full of eggs. In this situation the bees clearly prefer to use worker comb and appear to be very restrained as regards the drone comb.

The question of course arises as to why drone comb cells are rarely used, although they are made from the same material and have the same shape as worker comb cells, albeit somewhat larger.

There is a parallel to this in the field of biodynamic agriculture. In this farming method, spray preparations are used that are matured in cow horns. Rudolf Steiner showed that this preparation should be done using only a cow's horn and not a bull's, because the bull's horn does not have the properties available to the cow's, which belongs to the female animal. Experiments have confirmed this in practice. In view of this fact, it is easy to imagine that the worker cells have different properties to those of drone cells, even though they share the same basic shape. This is comparable to the situation with the cow and bull horns, although two very different animal species are involved. Bees appear capable of distinguishing the properties of different cells and therefore generally use worker cells for honey and pollen storage.

If bee colonies are located in an area where an early nectar flow occurs that is also the main flow upon which we depend for our harvest, then we must manage our colonies in order to take advantage of this early flow. To do that, we must adapt to the specific developmental stages of the bees.

Our goal must be to have as many flying bees in the colonies as possible. The bees which, as winter bees, are in colonies in spring, have become so weak that they are no longer capable of making the most of the available forage. In order to balance out the dip in the springtime colony population when the old bees dwindle, and thereby make full use of the early nectar flow, we cannot do without very deliberate stimulatory feeding.

Stimulatory feeding usually involves giving the bees honey-sugar fondant, thereby increasing the colony's inclination to produce more brood. After a certain time, through the resulting increase in laying, more flying bees become available. The difficulty is knowing roughly when the flow starts and therefore when to begin the stimulatory feeding.

The start of the flow can be determined approximately by observing over many years the site of the colonies and the development of the surrounding plants. Thus, if we know that dandelions flower at the beginning of May, and that the cosmic constellation of Mercury favours nectar at this time, we will start stimulatory feeding about 40 days beforehand. This is the length of time it takes, from egg laying, for workers to become flying bees.

So if, at the beginning of May, there are to be enough flying bees in our colonies to take full advantage of the dandelion flow, we would have to start stimulatory feeding on March 20. However, stimulatory feeding should be started only when there is already some foraging, and the first pollen from sources such as hazel, alder and willow is available. If at this time it is still too cold for the bees to fly, we will do the colonies more harm than good by feeding. The still present winter bees would overtax their powers in nursing the brood, because pollen, indispensable for increased brood care, is still in short supply. In this case it would be better to delay stimulatory feeding until the first pollen is evidently available.

This 40-day rhythm in bee development points to a very big difference between house bees – the part of the colony that tends to the hive – and the colony's army of flying bees. We get the sense that during this period, something specific is being prepared so that a new impulse can come into effect at the end of the 40 days.

During this period, the life of a worker bee unfolds. She completes her time as a house bee and after 40 days leaves the organism of the bee colony in order to become a flying or field bee. We might be tempted to think that flying bees are no longer in such close contact with the colony, but this would be to view the matter in a purely superficial way. As a house bee, the worker bee was responsible for the harmonious life of the colony. Now she is discharged from this responsibility and takes on a wholly new set of tasks. She is confronted with all the difficulties that face her in the outer world, but at the same time, through gathering water, pollen and nectar, she guarantees the basic nutritional substances for the colony living in the darkness of the hive. Thus, the worker bee frees herself by day from the direct sphere of influence of the queen and turns towards the outside world. Only in the time when she cannot fly does she return to this sphere of influence, but without being active in it like a house bee. The fraction of a bee's life spent as a flying bee represents an almost totally different life to that of a house bee, but it is nevertheless one that remains devoted to the service of the colony living in the hive.

If the preparations and stimulatory feeding have proceeded such that at the first flow there are enough flying bees, we can calmly look forward to the end of the flow. There just remains the hope that the cosmic constellations work so that the harvest compensates for the effort, and the colonies can bring in sufficient stores.

Nectar

Because various plants yield nectar at different times of the year, when foraging in a flow begins it is always exciting to discover which plants the bees are visiting. The easiest way to find out is to follow the bees' flight lines. A beekeeper who has kept bees in the same place for years will be able to recognise a known forage source from the particular direction of a flight line. These sources of forage can differ. They may be from flowers, or from trees yielding honeydew.

There are two kinds of nectar released by plants and they differ primarily in how they are produced. One kind of nectar comes from flower nectaries, derived from plant organs situated in the floral region of the plant. Here it makes no difference whether they are the flowers of herbaceous meadow plants or flowers of trees. The bees find flower nectar only in flowers.

The other type of nectar comes from dew nectaries. Dew nectar consists of substances similar to nectar that can arise whenever plants, primarily woody ones, produce fresh growth, for example leaves or needles. Various kinds of aphid are necessary for producing this dew nectar. This means that these dew nectaries do not come from the plant organs that produce it, but from an intermediary between plant and bee, namely an aphid that has to be present to suck out the plant juice and release it in such a way that it becomes available to the bee.

The purely external difference between these two types of nectar is therefore that flower nectar comes from specially provided organs, whereas dew nectar can only occur when the mediating aphid is present.

Those who are familiar with the subject will be surprised by the term 'dew nectar'; normally we use the terms 'forest honey' or 'honeydew'. As there is an enormous difference between the substance sucked up by the bee and the substance that we later harvest by centrifuging, the somewhat unusual term dew nectar has been very deliberately chosen.

Plants can supply flower nectar only as long as they are in flower. This does not mean that the first dandelion flowers make nectar available and keep this up until the last flower has faded. The capacity to supply nectar is very dependent on cosmic rhythms. It is these that cause the plant sap to flow upwards more or less strongly, or let the nectar organs produce more or less nectar, and weather patterns influence this flow as well. Soil conditions also affect the extent to which plant nectaries can produce nectar. It is clear therefore that many factors must work together for us to be able to count on the nectar release of plants.

How much the formation of plant nectar depends on cosmic rhythms can be observed anew each year. The world of plants may be the same every year in any given locality, the diversity may even be the same, but the nectar development is highly variable. It ranges from a real excess in one year to a very small yield in another. In most cases, the weather is regarded as responsible for these variations. However, taking into account the fact that the weather is to some extent dependent on cosmic rhythms, it is clear how all plant growth and nectar development proceeds in harmony with the cosmos.

A connection to cosmic rhythms can also be seen in the production of dew nectar. But the effects of these rhythms must now be considered not only in connection with the plant, but also as regards the aphids. If aphids are absent, the most perfect 'honey constellations' may occur, yet no dew nectar is produced. Then there are years when there is a surfeit of aphids, but the appropriate constellations that cause the necessary flow of plant sap are lacking. From this it is clear that for the production of dew nectar, several factors must work together for the sap flow in the plants to coincide with the presence of a sufficient number of aphids.

If these conditions are met, then the creation of dew nectaries proceeds as follows. The aphids, located on new spruce shoots or on the underside of oak leaves, insert their proboscis into the soft branch or leaf veins and suck up the phloem from the sieve vessel system. Phloem is vascular tissue that transports the soluble, organic compounds manufactured in photosynthesis to the parts of the plant that need them. Chief among them is sugar in the form of sucrose. The aphids filter out certain substances that they need to live on, partially transform them and excrete the residue. This excreted, somewhat altered, phloem sticks on the leaves or the needles of the branches below. As this substance contains sugar, it is licked up by the bees and carried into the hive as dew nectar. The term dew nectar came about because the excreted juice remains sticking to leaves or needles like dew drops.

The conversion of nectar to honey

Many people who enjoy eating honey do not know how this substance is made. They often think that the bees gather honey and carry it into the combs and then the beekeeper removes it. Considered superficially this sequence of thoughts is correct. But a close comparison of the substances gathered by the bees with those harvested from the comb shows that two rather different substances are involved: nectar and honey. To turn nectar into honey requires various transformations to take place. As the precise description of what happens between nectar and honey can fill a whole book, we shall deal here only with a few important aspects.

Before the bee sucks up the nectar she may dilute it with a little saliva (although this is not always necessary). By adding saliva, the transformation of nectar to honey has already begun. The bee now sucks up the nectar and as it passes through the pharynx into the honey crop, further glandular secretions are mixed with it. The honey crop serves as a sort of transport tank for the bee. The bee then flies from flower to flower until she has filled her crop, or there is no more nectar in the flowers. When she gets back to her colony, she gives the nectar to a house bee, who carries it to the cells and adds glandular secretions once again. Once the nectar is stored in the cells, it is further enriched by the bee's glandular secretions through repeated sucking up and discharging.

Nectar has a high water content of around 70 per cent, whereas in honey it is around 20 per cent. In order to reduce the nectar's high water content, the humidity in the honey chamber is kept low enough that the dry air absorbs the water. We can experience this process from outside the hive. If we visit our bees on the evening of a day when there has been a good nectar flow, we can see the bees ventilating vigorously. If we stand in front of the entrance, we will be enveloped by the wonderfully sweet scent of this nectar-honey. This arises from the constant exchange of air in the colony. Dry air replaces the air laden with nectar moisture, thus spreading a lovely smell round the apiary. Nectar also has lots of sucrose. In honey this is completely changed to fructose and glucose as a result of the bee's own glandular secretions. In this way, nectar is transformed into honey.

In order to discover when the nectar has ripened to honey, we take a honeycomb out of the colony and hold it horizontally so that a comb side with cell openings faces downwards. We then briefly shake the comb downwards. If honey drips out of it, it is not yet ripe. If nothing drips out of the comb, the honey is ripe and can be harvested. If we attach importance to high quality, we will try to leave the honeycombs in the colony until the cells are totally filled and capped. However, if melicitose honey appears (a form of honeydew containing polysaccharide sugars), we will not want to wait that long because by then it will not be possible to centrifuge it. Other honey types improve in quality through capping, acquiring a much better taste.

Rudolf Steiner said that bees were bred for the human being in order to turn nectar into honey by transforming it and enriching it through the influences of cosmic forces.

Honey is a very valuable substance for the human organism and although it is beyond the scope of this book to discuss the many trace substances it contains, it is worth noting that many of these can be regarded as pathogen inhibitors. They also occur to some extent in propolis. Through these substances, honey deviates somewhat from the category of food and comes closer to the category of a medicine. This does not mean that we should regard honey as a sort of medication, but that we can also use it as such. To treat it as a medicine, the honey has to be not only ripe as regards its material nature, but really matured. Maturing means the honey should all be capped. Only then will it properly meet the requirements of a medicine. Rudolf Steiner said that bees were bred for the human being in order to turn nectar into honey by transforming it and enriching it through the influences of cosmic forces. He indicated that this is extraordinarily important for human development. The effects of these forces are those that make honey into a unique and high-quality substance.

Rudolf Steiner frequently referred to the connection between the quartz crystal and the bee cell. He emphasised that a crystal that has points at each end has more properties than a crystal that has a point at one end only. If we draw a parallel between a quartz crystal and a bee cell, we will see that a capped cell, comparable with the crystal with two points, has different properties than an uncapped cell. This may clarify how important it is to obtain honey from capped cells, because it has taken up forces from the capped cell that are of a higher quality for human beings than those occurring in an uncapped cell.

FIGURE 40. *Quartz crystal with two points. The model of the bee cell.*

Therefore, although we can harvest ripe honey from uncapped cells, if it is to be properly mature honey it must be left until the cells are capped.

Processing the honey

Bees provide us with only high-quality honey. However, if we do not handle the honey with care, then it can suffer during the various stages of processing, from removing the honeycomb from the colony to bottling the honey. If the quality of the honey is therefore to remain stable, which it easily can for

FIGURES 41 AND 42.

Top and bottom: Honeycombs containing mature honey.

many years, we must take note of certain factors.

First, we delay comb removal until the majority of cells are capped. At warmth/fruit times or light/flower times we carry out the shaking test mentioned above to see if the honey in the uncapped cells is ripe. If this is the case, we place or hang all the honeycombs in a comb rack. Next, we brush the bees from the combs into the empty chamber and put the brushed honeycombs into either a closable comb transport box, an empty honey chamber or a super that can be closed with a floor and cover. The honeycombs should not be left long in the open, as this would incite bees to robbery, and the honey still in open cells may also absorb moisture from the humidity of the air, especially on rainy days (ripe honey has a water content of less than 20 per cent). This results in a deterioration in quality. Therefore, immediately after brushing off the bees, the honeycombs should be placed in a closable box for transportation.

The full transport boxes are carried as soon as possible to the extractor room for centrifugation. If the extractor room is far from the apiary, then it is helpful to heat the room so that the honeycombs that have cooled during transport can warm up again, enabling the honey to flow out of the cells more easily.

The extractor, honey-containers and strainer, as well as the uncapping tool, should be spotlessly clean. The honeycombs are now uncapped using the uncapping fork and placed in the centrifuge. The combs are then spun in the extractor until they are empty. Special care should be taken with white combs, because they can break easily if spun too fast. The honey now flows through a strainer into the settling or bottling tank. The strainer removes particles of wax that remain after uncapping the combs.

The honey obtained is left for three to five days in a closed container until all of the bubbles

FIGURES 43 AND 44.
Top: Full honey chamber.
Bottom: Comb transport box.

Figure 45. Honeycombs are uncapped before centrifuging. There are various methods. We prefer an uncapping fork with cranked tines.

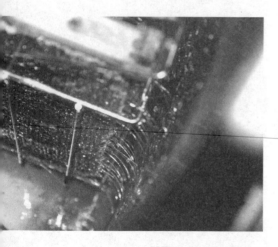

Figure 46. When centrifuging the combs, take care to put combs of the same weight opposite each other, otherwise the imbalance shakes the extractor and it starts to dance about.

and smallest wax particles have floated to the surface. This forms a foam that can then be removed by skimming. This is done in the morning and the evening. When no more foam reaches the surface, stirring the honey begins. Flower honey should always be stirred so that it crystallises out or granulates finely. A finely granulated honey can develop a better flavour than a coarsely granulated honey. For example, if we let rape or dandelion honey granulate without stirring, we will get very coarse crystals resulting in a poorer flavour. But if this honey is stirred well, it acquires a wonderfully mild taste and melts like butter on the tongue.

An electrically powered spiral or a triangular limetree or hornbeam wooden rod may be used for stirring. Always stir from the perimeter of the container inwards to the centre in a spiral pattern. The honey is stirred morning and evening for about five minutes. If a spiral stirrer is used, one or two minutes suffices for a 50 kg

(110 lb) container. If foam appears on the surface before stirring, it should be removed so as not to stir it into the honey. When a small groove or streak forms in the honey behind the stirrer and disappears only slowly, then we can stop stirring. Now the honey can be bottled.

FIGURE 47. *If possible, the honey should be strained straight after centrifuging. Freshly centrifuged honey passes more easily through the strainer.*

FIGURE 48. *Small air bubbles that get into the honey through centrifugation float up to the surface with small particles of wax. After two days you may start to remove the foam. For best results some practice in 'skimming' is advisable. The required tools are a spoon and a dough scraper.*

FIGURE 49. *Stir the honey so that it crystallises finely and evenly.*

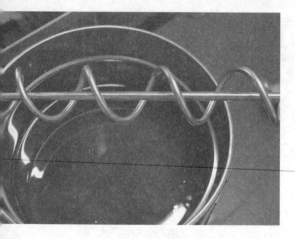

FIGURE 50. *A spiral stirrer used to stir the honey.*

If we have a flower honey that we do not wish to bottle immediately, but instead distribute in winter or in the following spring, the stirring should definitely be done soon after extracting. It has been discovered that coarsely granulated flower honey can no longer be returned to fine granulation through later stirring. As far as honeydew honey is concerned, stirring does not improve it, and as it remains liquid for a long time consumers are rarely offered granulated honeydew honey.

In view of the fact that stirring not only makes more finely granulated honey but also improves its taste, it is clear that stirring further refines the product of the bees. If possible, this should be done at light/flower or warmth/fruit times as these have a positive influence on the flavour of the honey.

In some honeys, plant-like shapes form during granulation. These formations, which sometimes look like froth, often down the side of the jar, are the result of the process of crystallisation. It is referred to as frosting. It does not reduce the honey's quality, but it arises in honeys of particular plant species. For example, if you have harvested pure rosebay willowherb (*Chamaenerion angustifolium*) honey, it may express this property so strongly that it seems as if there is pure honey foam in the jar. However, if such a honey is warmed and re-liquefied, almost no trace of these 'sugar patterns' remain.

FIGURE 51. *The jar should rest immediately below the tap so that as few air bubbles as possible enter the jar during filling.*

The bottled honey is checked again after a few days; any bubbles that have risen should be punctured and the jars or tubs securely closed. The closed containers are now stored in a cool, dark but dry room. It is important not to place jars of honey too long in direct sunlight because the sun's rays can cause changes in the honey, deteriorating its quality. Rooms in which honey is processed should not be lit with a large window but have a small window in order to eliminate damage to the honey from the sun's radiation.

In the chapter on combs, we discussed the sun-like qualities in the bee colony. In view of this, we might wonder why honey reacts so sensitively to the sun's radiation. If we look at the arrangement of the honey stores in the bee colony, we will see that they are included in the rounded bodily shape. The bee colony as a whole wants to live in darkness not in the light. This is clear from the fact that it always seeks a dark cavity in which to develop. It is above all the flying bees that are exposed to the sun's radiation, and it is because of this that they are somewhat separate from the colony which lives in darkness. The honey destined as food for the colony certainly absorbs the sun-like quality of the colony's bodily shape, but it is very susceptible to the sun's rays which can cause chemical changes. This tells us that the sun-like shape has properties other than a reaction resulting from direct radiation.

Types of honey and their uses

We have already discussed the two kinds of honey that occur naturally. One is the large group of flower honeys and the other the honeydew honey group. In regions where both kinds occur, people frequently argue about their relative merits. Honeydew honey is generally preferred. This is probably attributable to the fact that it remains liquid for longer and may be somewhat milder. People are even willing to pay more for honeydew honey than for the lighter-coloured flower honey. The darker honeydew honey is, the better it is. Nevertheless, keeping in mind the role of bees in the world, as well as the very different ways that flower honey and honeydew honey arise, we can only conclude that flower honey is preferable to honeydew honey.

Rudolf Steiner indicated that bees originated during the time of Atlantis. He explained that certain flower substances were needed for human development, and in order to obtain these substances the bee was created from the fig wasp, which is a purely flower creature (*Bees*, p. 102 and p. 179). The bee that resulted from this is an insect dependent on flowers for its life. From flowers she gathers the two basic nutriments, pollen and nectar, and the latter is changed into honey. Flower honey is therefore a substance of great importance to humans. It comes from flowers and contains the effects of forces imparted to it by the hexagonal cells.

It is clear from the foregoing that bees do not only make this flower honey for humans, but also for the bee organism itself. Bee colonies cannot live without flowers. The requirement for honey can be replaced by honeydew honey if necessary (although honeydew honey is not suitable for wintering), but nothing can substitute for flower pollen. Pollen that might in some years come from conifers is inadequate for the bees, not only in quantity but also in quality. Bees therefore need flowering plants for the growth and maintenance of their organism, but also so they can fulfil their mission for humans with the help of floral honey.

Rudolf Steiner made it clear that floral honey is extremely important for humankind (*Bees*, p. 51ff). What about honeydew honey? Can it be compared with floral honey at all, or is it just a substance similar to honey? Chemically it is equivalent to floral honey. We should not ignore the chemical differences between the individual kinds of honey, but should we ignore the differing origins of flower honey and honeydew honey as described above? Does it matter if nectar comes from the plant organ meant to produce it, or if it is only available via aphids? Rudolf Steiner even indicated that it is something special for the flowers when the bee drinks from them. He indicated that before drinking nectar the bee first secretes substances like poison, that are essential for the maintenance of the life of the plant world. Here we discover something highly interesting from the social point of view. The bee takes from the flowers the nourishment essential for its life only after it has given to the flowering plant substances essential to the plant's life. A harmonious balance arises between flower and bee that is supported by cosmic constellations, and by this means a varying abundance of nectar is produced in the plant organs.

Honeydew honey is unthinkable without the help of aphids.

It does not arise from the plant organs intended for it. Can it nevertheless be regarded as honey? Some might answer this in the affirmative, because the processes that turn dew nectar into honeydew honey are similar to that of floral honey. But it certainly cannot be maintained that honeydew honey is better than floral honey.

In the comparison of flower nectaries with dew nectaries we can see a parallel in water. The earth has organs from which water flows more or less strongly, depending on the cosmic conditions. We are referring here to springs, which may be compared with the nectar organs of the plant. But we can also tap into the flow of groundwater. This is comparable with the plant's phloem system. With the help of pumps and pipes this groundwater can be brought to the surface. This is comparable to the tapping of phloem by the aphid. Water from springs and water from boreholes are chemically the same, but do they have the same qualities for humans? People must find their own answer to this.

The impression may have been created that honeydew honey should be regarded as of lower value. However, we should not judge too harshly, because creation has enabled bees to harvest this sweet substance as well. Honey may only be regarded as being of poorer quality if, for example, its water content is too high, or it has been overheated. Judging the different types of honeys and their varieties on their material content should be avoided, because, from a dietetic point of view, different honeys have very different effects that can be used for specific purposes. Thus, a lime flower honey has effects different from those of a dandelion honey, or even from a honeydew honey.

Here are just a few of the dietetic effects of different kinds of honeys that we have found:

❀ Dandelion honey and sunflower honey support liver processes.

❀ Rape honey stimulates bile flow.

❀ Germander honey helps with bronchitis and asthma.

❀ Thyme honey and lavender honey strengthen the breathing.

❀ Lime flower honey, raspberry honey and cornflower honey strengthen the kidneys.

❀ Willow honey strengthens digestion.

❀ Hawthorn honey and sloe honey strengthen the heart.

❀ Honey from orchard flowers strengthens the vascular system.

❀ Honeydew honey can help with respiratory conditions.

In general, honey is a very useful substance that can help us strengthen our own organism.

Chapter 10
Feeding in Winter

During the summer, bees gather nectar and turn it into honey so that they can not only give it to us, but also ensure that they have sufficient stores for winter. Bees need winter food stores because, contrary to what people in general think, bees do not hibernate, they remain active to some extent. This can be observed purely acoustically from the constant, deep humming of a colony in winter. The necessary quantity of feed for a colony depends not only on its strength, but also on the weather. If the weather stays unchanged throughout the winter, food consumption is usually less than when the temperature is constantly oscillating, which prevents the colony from taking its winter rest. In Central Europe we can say that for every comb occupied by bees, a colony should have around 1–1½ kg (2–3½ lb) of winter stores. Thus, a colony that occupies ten combs in the autumn should have more than 10–15 kg (20–35 lb) of winter food.

Because the honey is harvested from the colony in summer, a substitute must be given for the winter. Since the beginning of the last century this substitute has been sugar. The fact that sugar feeding is just a substitute repeatedly sparks heated discussions.

Beekeepers have long since become used to feeding sugar, but honey consumers remain very sceptical because they assume that some of the sugar used for winter feed finds its way into honey. This scepticism goes so far that some will not eat any crystallised floral honey because they assume that it is made of conventional sugar. Because floral honey usually sets after some weeks, many people do not get to know their local floral honey and therefore cannot form an opinion on the taste difference between floral and honeydew honeys. The result of this unreasonable, factually wrong, assumption is that liquid honey in the form of dark honeydew honey is preferred.

Many nutritionally aware people demand that beekeepers stop feeding sugar and let the bees winter once more on honey, or at least on raw sugar. Whilst feeding raw sugar is a nice thought, it does not work in practice. It contains so much molasses that the bees cannot filter it out. The molasses would so rapidly fill the bee's rectum that the bee would have to defecate in the middle of winter. This would then take place in the colony. The instinct of the bees for cleanliness urges them to lick the faeces off the combs. In doing so, they would take in a bee pathogen that can cause diarrhoea. This clearly shows that raw sugar is inappropriate for winter feeding.

The fact that in earlier times bees wintered on honey is attributable to some factors which we now only find very rarely in beekeeping. In order to get to know these factors better, let us look at the course of the year in a skep (or basket) of the sort that was kept on farms everywhere in Germany before 1900.

The bees kept in those days were the native European dark bees (*Apis mellifera mellifera*). They developed only very small colonies with a correspondingly small brood nest. The skeps used were proportionately smaller than our modern hives. Rearing of the brood began in spring. A girdle of pollen was deposited around the brood nest and the rest of the space was filled with spring flower honey. The skep had a bunghole on top into which a wooden bung was inserted. The beekeeper removed the wooden bung to see if the colony needed more space. If the bees had started to build small combs on the bung, they would know it was time to remove the bung and put a small basket on the top of the skep to serve as a super, or honey chamber.

The colony now filled this little basket – called a cap in some places – with comb and filled the comb with honey. So now in the brood chamber skep we have the brood, the pollen arch and the honey arch. In the honey chamber (or cap), we have only honey. This honey was the harvest for the current year. In good years the beekeeper could cut out and harvest the cap's contents several times. If a dew nectar flow started, this was guaranteed to be stored in the cap, because in the skep below there was not enough space available. When the time came that no more honey was expected, the cap was removed and the bunghole again closed with the wooden bung.

The spring flower honey in the skep at the bottom was left entirely for the colony. This was the colony's winter food. In the following year, on the Thursday before Easter, a quarter of the combs would be cut out. This resulted in all the combs in such a skep being renewed within a period of four years. The process began again with the beekeeper checking the wooden bung.

In this context, we should mention that not all floral honeys are equally good for overwintering. Heather honey is not suitable, neither is monofloral rape honey. It seems that the only floral honeys suitable for overwintering are those that are also available in the spring, such as willow, dandelion and orchard tree flowers. From this we can assume that honey for overwintering should be as versatile as possible, because then it has a greater value for the bees and does not lead to one-sidedness.

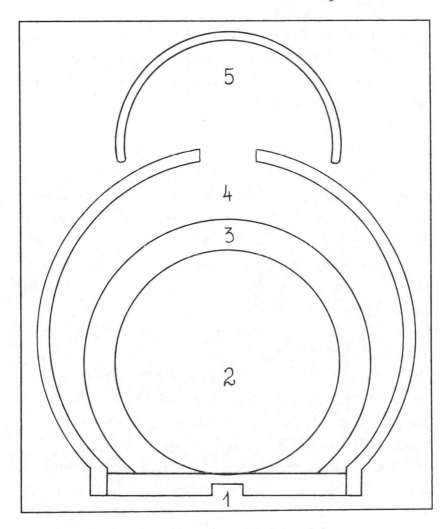

FIGURE 52. Arrangement of stores in a traditional skep.
1. Entrance.
2. Brood nest.
3. Pollen arch.
4. Arch of spring floral honey for winter food.
5. Cap with summer forage honey that is harvested by the beekeeper.

Skeps have largely disappeared from modern beekeeping, instead we have our square or rectangular boxes. These boxes contain movable combs that are big enough to provide enough space for the brood nest and the pollen girdle. But we want to have the honey stored on separate combs above the brood nest because that makes it easier for us to harvest. We now have bee races that develop significantly larger colonies and brood nests, and therefore our bees cannot be kept in such small skeps. They would swarm themselves to death. Meanwhile, the availability of flowers has become significantly reduced. In some areas there is no longer a main flow of flower nectar, so beekeepers must focus their attention on the honeydew flow. But bees can winter as little on honeydew honey as on raw sugar. This comparison with skep beekeeping shows that overwintering on honey is hardly possible any more, and only works in places where enough floral honey is available for feeding to the bees during winter.

Because bees can convert sugar to a substance that is chemically similar to honey, sugar has prevailed as the winter feed of choice. It is the beekeeper's job to give the sugar at the appropriate time and without it costing the colony too much effort. The beekeeper can also add teas to the sugar in order to supply the floral forces that promote bee health.

Winter feeding should be started at light/flower or at warmth/ fruit times after the last honey harvest. The colonies are still strong at this time and have many bees that will not go into winter because they are too old. As the transformation of cane sugar to winter feed suitable for the bees uses up their energies, this time is opportune because the old bees can then participate in the work. We must also keep in mind that winter feed given promptly is transformed more readily than feed that is only given to the bees when they can no longer fly outside. Furthermore, prompt feeding can make the bees raise more brood, which likewise can be a positive influence.

The honey chambers are removed and the colony is given the quantity of combs on which it will winter. The brood is inspected again, and the honey stores in the brood chamber estimated. The colony should be left with at least 2½ kg (5½ lb) of honey. As a guide to estimating honey stores, we can say that a hand-sized area of comb capped on both sides comes to half a kilogramme (1 lb). The deficit in stores is now made up by liquid feed. Feeding should begin in the evening of the day on which the honey was removed. The colonies will be grateful for this because the removal of the honey represents a great loss for them, which can only be compensated for by immediate feeding.

The winter feed is prepared in the ratio of 2:1, that is 2 kg of sugar to 1 litre of water (4 lb to 1 quart). A litre of this sugar solution gives about 750 g of prepared winter feed (1 quart gives 11/2 lb). For every 100 litres of this syrup add 2–3 litres of tea (2–3 quarts to 25 US gal). Rudolf Steiner recommended adding chamomile tea to the winter feed. He indicated that sugar lacks the floral forces that the bees need, and that chamomile, because it cannot contribute to making honey, has such concentrated floral forces that it can impart them to the sugar solution via tea. He also advised mixing a pinch of salt in the winter feed, as this makes it more digestible for the bees (*Bees*, pp. 38f).

Supplementing with teas serves to strengthen the organism of the individual bee, so that it is less susceptible to disease.

Because chamomile does not help fight all of the diseases bees face today, we have started adding teas made from plants used in biodynamic preparations to the winter feed. We have had the best results with these teas. Supplementing with teas serves to strengthen the organism of the individual bee, so that it is less susceptible to disease. Up to now, after keeping bees for over 50 years, we still do not need to give them medication of any kind for curing disease. To make the teas we use yarrow, chamomile, dandelion and valerian. We take the flowers from them and make an infusion with boiling water, leaving it to draw for 15 minutes before straining it. In addition, nettle, oak bark and horsetail are placed in cold water, briefly boiled, and then strained after 10 minutes. This tea is added, together with the strained floral tea, to the winter feed which comprises half a litre (half a quart) of tea from each plant, making altogether 3½ litres per 100 litres (3½ quarts per 25 gallons) of sugar solution. For this amount of winter feed we use about 5 g (just under ¼ oz) of each individual preparation plant. However, the teas from the different plants are prepared separately each time.

In the course of the year anyone may collect for themselves the plant parts needed for the teas. Some examples are given below. The plant parts should be spread on paper and dried in a shady, ventilated place.

❋ Yarrow (*Achillea millefolium*) is collected at warmth/fruit times when the sun is in the constellation Leo.

❋ Chamomile (*Chamomilla officinalis*) is collected at light/flower times, shortly before June 21.

❋ Dandelion (*Taraxacum*) is collected early in the morning at light/flower times. The flowers should still be closed in the middle and therefore not yet pollinated, otherwise they become 'dandelion clocks' in the drying process.

❋ Valerian (*Valeriana officinalis*) is collected at light/flower times at the time of the festival of St. John.

❋ Nettle (*Urtica dioica*) is collected at light/flower times. They should have the flower primordia, buds, at the end of stems from which the flowers will grow. The whole above-ground part of the plant is used.

❋ Oak bark (*Quercus robur*) is collected at earth/root times. Pieces of oak bark – no bast, or inner bark, as it contains too much tannic acid – are rubbed on a kitchen grater.

❋ Horsetail (*Equisetum arvense*) is collected at light/flower times. The above-ground part of the plant is collected.

Chapter 11
Bee Diseases

Considering the substances that bees consume, it is almost incomprehensible that they can become diseased at all. Honey and pollen are considered healthy, disease-resisting substances by humans. So how does it happen that bees repeatedly suffer from diseases that sometimes lead to the death of whole colonies? A simple answer to this might be that the bees are so weakened in their organism that any pathogen can easily gain a foothold. Another contributing factor arises from problematic food substances which, as a result of constipation and diarrhoea, cause bees considerable difficulties.

If bees were kept and raised in a bee-friendly manner that supports the whole colony organism, we would not be worrying today about how we will keep bees healthy for our descendants.

In trying to discover why bees are so prone to disease nowadays, we do not have to look far for the ultimate cause: the human being. If bees were kept and raised in a bee-friendly manner that supports the whole colony organism, we would not be worrying today about how we will keep bees healthy for our descendants.It is not possible here to mention the many causes that have contributed to the increasing susceptibility of bees to disease, but we shall mention a few.

❋ In the section on combs, the round, sun-like shape of the colony's body was discussed. This shape and the effects of its forces are ignored in modern beekeeping's square or rectangular boxes.

❋ The effects of forces that are attributable to cell shapes get mixed up through so-called artificial queen breeding.

❋ By introducing alien, already mated queens, colonies are deprived of the process of rearing their own, new queen.

❋ Using foundation provides the bees with a poorer quality of material to construct their combs.

❋ Feeding sugar as winter feed gives them a honey substitute that lacks the floral forces necessary for life.

❋ By making hives with glued wooden boards, such as plywood or chipboard, or with foams, we create an unnatural screening effect against influences from the surroundings.

❋ By breeding and keeping non-local bee races, we deprive those races of the pollen and nectar types to which they had originally adapted.

- Through flying for forage in totally different landscapes, the bee organism repeatedly has to readapt itself to them.

- Through management methods involving, for example, frequent shuffling of the brood frames, or moving the entrance for swarm control, we disturb the integrity of the colony organism.

These points are unlikely to be recognised by all contemporary bee scientists, but whoever wants to get closer to the causes of disease in our bees should give these points careful thought. Of course, purely external influences, for example the decline in floral diversity, should not be ignored, but as we cannot influence such factors ourselves, I will not discuss them further.

Bees are prone to a number of diseases that, broadly speaking, fall into three groups: brood diseases, adult bee diseases and diseases that harm both brood and adult bee.

Brood disease

Malignant foulbrood

Malignant foulbrood, also known as American foulbrood, is caused by the pathogen *Paenibacillus larvae*, a rod bacterium. The stable form of this bacterium is a spore that gets into the larval gut via its food. As soon as a round larva develops into a straight larva, the active form of the bacterium, the rod, grows from the spore. Because of the favourable conditions in the larva, the rod reproduces so rapidly that, within one or two days, it destroys the whole larva and the larval skin. This destroyed larva becomes coffee-brown and sticks to the cell wall.

A colony infected by foulbrood is recognisable because its brood cells do not hatch. The cell cap sinks somewhat and usually has a hole in the middle. Some bees try to remove the cap in order to discard the destroyed larva, but they do not succeed because the larva is tough and sticky like rubber. If we suspect the presence of foulbrood, even if no sunken cell caps with holes are present, on close inspection we might see larval plaque or residue. To confirm our suspicions, we can dip a matchstick into this residue and if, when we take it out again, a rubbery thread forms, then we can assume that the colony has foulbrood.

So how did foulbrood get into the apiary? There are countless opportunities for infection. The bees might have licked out a honey container at a rubbish dump on which there was contaminated honey. An infected swarm could have moved into an empty hive, or we might have inadvertently taken and hived an infected swarm.

Spreading the infection at the apiary is triggered either by the bees drifting or during our own hive inspections. For example, the bees may have some bacteria on their body hairs that they transfer to another colony through drifting. The bees whose task it is to clean then infect themselves with the bacteria. Later they feed the brood and thus the bacteria get into the larva in its food. This colony has therefore been infected by a flying bee. If we inspect an infected colony and then an uninfected one, we can likewise transfer the bacteria and in this way infect the entire apiary.

The beekeeping instructor, Ernst Perkiewicz, vividly illustrated this infection route frequently in his lectures and courses. He showed that we should imagine infected hands as if coated with a red paint. As soon as anyone with red hands touched something, whether a hive tool or overalls, or even the hive of the neighbouring colony, more and more red marks would be left behind. This striking example of the spread of foulbrood spores shows how essential it is, once foulbrood has been diagnosed, to be careful of what we touch so as to avoid contaminating everything in the bee house, the extraction room or even our car if we use one. If the foulbrood has been eradicated in the apiary, but there are still spores in the extraction room or our car, then the eradication will not last.

Paenibacillus larvae is a very resistant bacterium. In order to kill it, it must be heated to over 120°C (250°F) for more than half an hour. It is said that it can easily survive for more than 40 years in old, empty hives. This indicates how cautiously and conscientiously we must proceed when buying and populating old hives, or when infected colonies are to be cured.

As foulbrood regulations differ greatly in different parts of the world, a general method of eradication cannot be given here. However, where the law allows, you may use the most humane and bee-friendly method. This involves artificial swarming.

Treatment

The following procedure is best carried out at light/flower or warmth/fruit times. In the afternoon, after bee flight has largely stopped, the entire infected colony is placed on a comb rack. The bees fill themselves with honey and, as described in the section on the artificial swarm, are brushed into the swarm box through a sheet metal funnel. Two or three brood combs are placed in the empty hive on which the returning flying bees can still settle. Then the hive is closed up again.

The swarm box is now placed in a cool, dry, dark room, where it remains for 48 hours. The swarm is allowed to get hungry. During this time, it is left entirely alone as constant disturbance (for example, switching on the light) would cause the swarm to use up its food too quickly and thus suffer harm from hunger. Some suggest giving the artificial swarm a litre (quart) of feed at this point. However, this is not necessary if, before the bees were brushed off the combs, they were able to fill themselves and are not subsequently disturbed. The bees use up their honey and groom each other. Through the grooming, the residues of food and foulbrood spores on their body hairs enter their intestines where they are destroyed. Therefore, during this 48-hour period of starvation, the bees cure themselves with the help of their instinct for cleanliness as expressed in their allogrooming.

In the evening following the making of the artificial swarm, the hive it was taken from is closed up and the bees that remain in it killed with sulphur. In doing this, not a single bee should be allowed to fly away. This is very important, which is why it is done just before it gets dark.

It must be assumed that all the combs of the colony are contaminated. Therefore, they should be burned in a deep hole in the ground, and afterwards covered up again with the removed soil so that no bee can get at any of the residues. Because it is not possible to say when foulbrood was brought to the apiary, it is recommended that the combs still in the comb store should be cut out and identified as contaminated wax to be sent to an appropriate wax rendering company.

The hive is scraped out and cleaned. To be sure of killing foulbrood spores, the whole hive and its individual parts, including the entrance, alighting board and hive front, should be scorched with a gas torch. To be sure that the scorching is intensive enough, the flame is moved slowly over the wood until the wood is discoloured. The hive tool and everything that has been handled must likewise be decontaminated, either by scorching or with a strong cleaning agent supplied for this purpose.

The hive is now equipped with new frames containing foundation made from sterilised wax or starter-strips.

When the 48 hours have elapsed, the swarm may be allowed

to run into the newly prepared hive. This should be the hive that the swarm lived in before the colony was treated, as otherwise unimaginable drifting could occur; the artificial swarm, like a natural swarm, does not forget where it is. To avoid any contaminated debris on the swarm box floor getting into the hive, the box should not be shaken empty. The swarm box must be decontaminated in the same way as the hive. After all these tasks, it is advisable that we wash our hands to minimise spreading foulbrood again. This colony is now treated as a swarm.

If there are several colonies in the apiary, the procedure needs doing with all of them. Because this cannot always be carried out in single day, it has to be continued on the day after. This means that for at least one day none of the apiary's hives should contain bees. Just one bee still flying around could start a new contagion. In such work we must be somewhat cruel and, in order to avoid the slightest risk, squash any bee that is moving about outside the artificial swarm box. If there are many colonies in the apiary, we must begin this work early in the morning as there will not be enough time if we leave it until the afternoon to begin.

Benign foulbrood

Benign foulbrood, also known as European foulbrood, is caused by different bacteria and is spread in a similar way to American foulbrood, but it is significantly less contagious. The larval residue does not pull into rubbery threads, although it may smell slightly sour. Colonies can rid themselves of slight infections, although with more serious infections certain approved veterinary medicines should be used. The artificial swarm treatment is also used here, and the entire cure and decontamination procedure is the same as for American foulbrood.

Chalkbrood

Chalkbrood is a fungal disease caused by the mould *Ascosphaera apis*. The spores enter the larva in their food. At the stage of the stretched larva or pupa, the fungal spores grow into fine fungal threads that spread through the larval body and dry out. White-grey stretched larvae or pupae remain in the cells and the bees try to pull them out.

FIGURE 53. *Chalkbrood larva.*

Its spread in the bee colony is due to individual bees, or even to slight draughts when combs are removed for inspection. There are always a few spores in a bee colony. Although there are always a few spores in a hive, it normally manifests as a disease when an imbalance occurs in the relationship between warmth and humidity in the bee colony. Chalkbrood can also arise when

FIGURE 54. *Chalkbrood larvae in front of a bee house.*

the colony's harmony is disturbed, as described in the section on swarm prevention.

As regards small infections that usually occur in spring, colonies can be rid of chalkbrood by removing the chalk larvae or pupae. If there is a more serious infection, for example during a wet spring, the infected combs should be removed and replaced with empty, white combs. In the worst case, all the combs must be exchanged.

Stonebrood

Stonebrood is a fungal disease caused by the mould *Aspergillus flavus*. It is very rare, but in already weakened colonies it can be harmful to adult bees. Its spread and outcome are similar to chalkbrood. If stonebrood spreads more seriously, it is advisable to kill strongly infected hives. Stonebrood is recognisable by dead, hardened pupae that take on a yellowish-green colour. The spores are also dangerous for humans because they can lead to unpleasant inflammations in the mucus membranes. We must take care not to get the spores in our eyes when inspecting combs.

Adult bee diseases

Dysentery

Dysentery is a kind of diarrhoea that happens when colonies overwinter on the wrong winter feed and the weather does not allow them to make an interim clearing flight. Raw sugar, honeydew honey, heather honey, as well as sugar solutions that have too much tea added, are not suitable as winter feed because they contain too many impurities (for example, molasses, that give rise to a premature filling of the rectum). As a result, the bee defecates in the hive. Its instinct for cleanliness nevertheless continues and it licks up the faeces again, resulting in permanent diarrhoea or dysentery. At the clearing flight, the colonies with dysentery are recognisable by the fact that they already defecate on the alighting board and sometimes coat the whole hive front with faeces. At the first colony inspection, the colonies with dysentery, which are often weakened, should be supplied with clean combs, narrowed down and kept warm.

Dysentery can also be caused by outside influences. Mice gnawing at the combs, birds pecking, or even the presence of the beekeeper can disturb the peace in the apiary so much so that colonies become agitated. This disturbs the quiet, steady consumption of food, thereby causing digestive upsets that may result in dysentery.

Spring sickness

Spring sickness happens in May and is recognisable by many young bees with swollen abdomens walking round in front of the hive unable to fly. If you carefully press the abdomen of one of them, a thin stool comes out. It is assumed that this digestive disturbance arises from a shortage of water in the colony, which may prevent the pollen from being properly digested. In such cases it is advisable to provide colonies with a water supply, or with dilute sugar solution in the hive. The water shortage is attributable to bad weather preceded by vigorous pollen foraging.

Tracheal mite

The tracheal mite, *Acarapis woodi*, also called an internal mite, lives and reproduces in the bee's anterior pair of tracheae through which the bee exhales. The tracheae are not closed with valves like those through which the bee inhales, but instead are always open. To find the tracheal entrance, the mite needs only the steady air stream that arises from exhalation. The mite has a proboscis that it inserts through the tracheal wall in order to reach the haemolymph (blood) system on which it feeds. It lays its eggs in the tracheae from which male and female mites emerge and then mate. Several generations of the mite may be found in the bee's trachea, and it is the presence of such large numbers of mites that causes the bee breathing difficulties. In addition, every time a mite penetrates the tracheal wall, it causes injury around which scabs form. These scabs further constrict the passage of air.

The tracheae are made of chitin, a fibrous substance that is also a major component of the bee's exoskeleton, and because it cannot heal, the pathogens may pass through the wounds into the bee's bloodstream. Seriously affected bees thus become weak and incapable of flight. The mite spreads by transferring from one bee to another bee next to it. The mite may spread between apiaries through drone-drifting or swarming, as well as through moving colonies. It is not spread in empty hives or on beekeepers, because without bee haemolymph it lives only a few hours.

Mite-infested colonies are detectable in winter by bees frequently leaving the hive, and in spring there is often evidence of dysentery. Furthermore, such colonies tend to raise brood in winter, which in spring shows as a very early departure of winter bees. As the tracheal mite only occurs on adult workers, queens and drones, the brood and newly hatched workers are free of it. The mite may be kept within tolerable limits simply by favourable spring and autumn foraging conditions that ensure a good supply of young bees. In localities where forage is sparse or completely absent, the mite may spread so fast that colonies collapse after one year of infestation. A cure is possible only with medication.

Nosema

Nosema is an intestinal disease caused by a single-celled microsporidian called *Nosema apis*. It enters the intestine by way of food and drink. There it matures, infects intestinal cells, multiplies and destroys them. This makes the bee unable to digest food; it weakens, becomes unable to feed the brood and dies. With more serious nosema infections, dysentery symptoms occur due to the destruction of intestinal cells. In spring, bees may be found crawling on the ground in the apiary unable to fly.

Nosema spreads rapidly during cool, wet springs in which winter bees, due to pollen shortage through their own excessive consumption of protein, have to look after and feed the brood. Through their over-expenditure of energy, the winter bees not only become weaker but also die earlier. Nosema finds its most favourable growing conditions in weak colonies. The severity of nosema diminishes only when the time of fruit blossom is over and colonies have achieved the transition from winter to summer. As regards prevention, there are some medicines that destroy only the active form of the pathogen, not its durable form, the spore. It can be assumed that all colonies have the spores.

Whether or not the spore finds an adequate breeding ground to become active is not just dependent on the spring weather. The method of colony management itself can also inhibit or encourage nosema. Furthermore, there are bee breeds that are particularly prone to the disease. As a preventive, supplementing the feed with chamomile strengthens the bee's intestinal organisation, but a general cure is available only through chemical medicines that keep the nosema infection and its spread within bounds.

As nosema is attributable to countless factors that differ greatly from apiary to apiary, we can only control it by helping the bees become stronger. Old combs or badly prepared winter feed can promote nosema, and wet places can be as much a stimulus to nosema as changing queens in the autumn. The example of queen changing shows how sensitively the bee colony organism can react.

Each spring, until the 1960s, the Marburg Institute for Bee Breeding and Research monitored the winter loss rate of individual colonies to nosema. These investigations showed that colonies that had received a young, bred queen in the autumn had an average to strong nosema infection. But colonies that went through supersedure at around the same time showed no nosema infection. This illustrates how differently colonies can react to queens that have not been reared in their own colony compared with queens who were able to spend their entire development in the colony to which they belonged. When we take into consideration how many queens are changed by beekeepers in the autumn, it is not surprising that nosema has become an ever-present side effect of beekeeping. These examples show how sensitively the bee reacts to certain interventions and makes clear how difficult it is to suggest a general healing or preventative measure.

Diseases that harm both brood and adult bee

Varroosis

Varroosis is caused by the mite *Varroa destructor*. The varroa mite lives on the haemolymph of the adult bee, larva and pupa.* It is extremely difficult to eradicate because it does not infest just the brood, or just the adult bee. Living on the adult bee, the mite punctures the chitin that forms the bee's exoskeleton at places where there are moveable parts in the armour. The chitin is particularly soft at these points so the mite can easily insert its proboscis in order to reach the haemolymph on which it lives. The spreading of the mite in the colony happens, among other reasons, through transference from bee to bee. It is very nimble and can even jump. Therefore, unlike with the tracheal mite, it does not depend for its spread on the bees standing very close to one another for a long time.

When it wants to lay its eggs, the mite searches for the brood. In the time when the mite began to get a hold in Europe, it was

* Very recent research has shown that the varroa mite feeds not on the haemolymph of the bee, as has been the accepted view, but on its fat body tissue.

assumed that it would only infest drone brood. But since then it has been realised that it also infests worker brood. The female mite that is ready to lay enters the brood cell shortly before capping. She lays her eggs on the larva. Male and female mites hatch from the eggs and puncture the larval skin to feed on the larval or pupal haemolymph. During the period when the cell is capped, the mites mate and, now ready to lay, emerge from the cell when the young worker or drone hatches. They then feed on the haemolymph of adult bees, whether queen, worker or drone, or they re-infest new brood cells and thereby continue the reproductive cycle.

Colonies are weakened by the varroa mite because the bees lose haemolymph through the puncture wound in the chitin, which also allows pathogens to enter their circulation. Brood infested by the mite mostly do not develop into healthy bees. Hatching bees that show deformities are not tolerated by the other workers and are thrown out. These bees often cannot fly and have stunted abdomens or wing buds that are no longer capable of developing into whole wings. If worker brood is badly infested, it is easy to imagine that this greatly impairs colony growth because each deformed bee can no longer make its contribution.

As the mite infests the brood and the adult bee, the success of any eradication method is relatively small. Up to now all treatments have to be given in the early spring, or in late autumn when colonies have no brood, because the effects of these treatments do not reach the brood through the cell caps. There are various substances derived from formic acid for killing mites, and whilst these substances differ in their application they are often used in their gaseous state. So far, formic acid has proved to be the safest treatment. Medicines will probably be developed that change haemolymph in such a way that the mite loses interest in it or it kills them.

In biodynamic agriculture Rudolf Steiner devised a way of controlling pest infestations. This method enables people to keep pest numbers so low that they no longer occur as pests but instead as isolated cases incapable of causing much damage. This method does not eliminate the mite entirely.

Rudolf Steiner indicated that when particular cosmic constellations occur, the whole pest or its parts should be incinerated on a wood fire and the ash from it spread as a fine pepper in those places where the pest should no longer be active (*Agriculture*, pp. 121f).

As there is often not enough ash, we embarked on experiments to potentise it homeopathically in order to transfer the effect of the ash to water as a carrier. Potentisation turns a small quantity of ash into a large volume of potentised liquid that can be applied by spraying over large areas. In our experiments, decimal potencies from D1 to D36 were prepared in order to find out which potency was most effective in restricting varroa reproduction. This involved diluting the ash in liquid by a factor of ten and then diluting the resulting solution by another factor of ten and so on, up to the required degree of potency. D8 gave the best results in experiments conducted over many years.

As Rudolf Steiner suggested only a few constellations for ashing, Maria Thun undertook a long series of experiments to discover which constellations gave the desired effects on weed growth or its decline. These experiments were spread over many years and prior to her death in 2012, Maria Thun was in a position to state the precise constellations for ashing weeds and animal pests. The effect of D8 on inhibiting reproduction is so clearly demonstrable that we can recommend this method to beekeepers for controlling varroosis.

It has already been mentioned that during the use of ash, no total eradication of pests and weeds is possible. This suggested to us that although it gives us the ability to keep weeds and pests within tolerable limits over a long period of time, it was also clear that we should use this time to discover the real causes and eliminate them.

Modern varroa treatments are reminiscent of some people who have a headache or toothache: they go to the pharmacist, buy medicine that relieves the pain, and take it repeatedly each time the pain recurs. Other people go to their doctor and show them where it hurts. They then try to diagnose what is causing the pain and prescribe a medicine for the transitional period until the cause has gone. But varroosis is always treated in the manner described in the first example. People believe they can beat the parasite with medicine of some kind or other.

For example, when we study how plants or animals acquire pests that live on their sap or blood, it is clear that it occurs when they are given fertiliser or feed that is completely wrong for them, or there are irregularities in their food substances.

Rudolf Steiner's indications on bee nutrition are therefore very revealing. He prepared beekeepers for a future when there would be a weakening in the bee's haemolymph system. He said this weakening is due to there being years when the weather prevents the bees from visiting flowering plants for the nectar and pollen that they need at exactly this time, and thus they may only provide themselves with honeydew honey. He recommended cultivating the corresponding seasonally dependent plants, so that they would provide the bees with flowers at the right time. This would be possible at specially sheltered locations set aside for this purpose. Furthermore, he indicated that it is a waste of time cultivating plants that give lots of nectar and pollen as such plants are capable of only an irritant effect. He suggested that beekeepers could set up small greenhouses in which they could cultivate plants that the bees definitely need, but which they might not be able to visit due to weather conditions (*Bees*, p. 87). We might add to this that we should cultivate not only the plants that are unavailable due to weather, but also those that have disappeared for other reasons, such as eradication.

If the bee is short of the necessary pollen at a particular time, a weakening of the haemolymph may occur. A more detailed examination and consideration reveals the cause of this weakness. As regards the bees kept nowadays, not only in Europe but also over the whole world, it is clear that the native races are absent from many regions. They have been suppressed by deliberate breeding of other races, or racially diluted as a result of individual beekeepers or beekeeper associations keeping other races. Even general apiological science contributed to this dilution. What does this look like in detail? Before the Second World War the northern bee (*Apis mellifera mellifera*) was the commonest in Germany. After the war, increasingly intensive breeding of the Carniolan bee (*Apis mellifera carnica*) started, because this gave better harvests and was less inclined to sting. The northern bee was our traditional native bee. The Carniolan bee was brought over the Alps from the Southern Austria-Slovenia region and introduced into a climate alien to it. The same thing happened to European and African bees that were taken to South America.

Individual beekeepers and beekeeper associations who were not satisfied with the bees they already had brought bees from other countries and crossed them with, for example, their native bee in order to get better harvests. The Buckfast bee of Brother Adam arose in this way.

Apiology institutes the world over, by keeping bees alien to the region, also contributed to diluting the native bee with their drones, thus creating fantastic mongrels. To give another example, in 1965 research experiments were conducted at almost all bee institutes in West Germany. They kept and tested various races and lines of bees from Eastern Europe, thereby contaminating the regions around their individual research apiaries with their drones. Considering how far drones and queens can fly to drone congregation areas, it is easy to imagine the implications of those experiments for the surrounding beekeepers and their bees.

With these few examples we can readily appreciate that, at least in the developed countries where there are beekeepers, there is hardly a single, racially pure, native bee. Wherever we look, alien races are tried out and crossed. This means that if we use a bee that is alien to a particular locality, we can never provide at the appropriate time the pollen it would find in its country or region of origin. For example, in mid-May in Denmark, Italian bees (*Apis mellifera ligustica*) cannot find the pollen that they would gather at this time south of the Alps. In South America, European bees do not have the pollen available at the appropriate time that they would have in their European country of origin. Especially with this last example, a reversal of the seasons applies. Therefore, we can say that as soon as bees are transferred to an alien locality, in which a different plant flora predominates, weakness must occur in the haemolymph through a shortage of pollen species, as indicated by Rudolf Steiner. This weakening of the haemolymph creates the best growth conditions for parasites like tracheal and varroa mites. Even if in future we find medicines to fight these parasites, new parasites will repeatedly arise until we have managed to re-establish native bees in their individual regions. Without doubt this will be a challenging process. Even so, this worthwhile goal is only achievable by breeding or back-crossing, because in many regions the few surviving native bees have long been diluted or crossed with alien bees.

This list of diseases contains only the most important and common ones. Likewise, from the listed diseases, only those are described that we may find at any time, and which we should learn to recognise.

When we look at bee diseases and try to understand that they are due to weaknesses of the bees, we should do all we can to eliminate these weaknesses. We should not seek to blame others or the environment, but should examine our own attitude and behaviour towards the bees. If we try to recognise the problems that we create for the bees through queen breeding, foundation, winter feed, hives and lack of landscape-typical plants, it would certainly be possible to gradually eliminate the factors that lead to weaknesses in the bees. But so that we do not lose heart, we must proceed step-by-step, because trying to control all factors at once would be beyond the powers of any individual.

Chapter 12
Methods of Ash Usage

As has already been mentioned, Rudolf Steiner recommended incinerating weeds and animal pests during appropriate cosmic constellations and spreading the ash like pepper as a way of controlling them. Because this method, in combination with potentisation, has given the best results in combatting other animal pests, I have in the past recommended it to beekeepers for controlling varroosis. After we had used this treatment for many years with our colonies, it appeared that varroa did indeed decline, but the colonies resented being sprayed with the liquid D8 potency: they became more easily inclined to sting. After thinking this over for a long while, we decided to use the varroa ash without potentising it by sprinkling it in the colonies. The results were good.

Unfortunately, we get very little ash by ashing mites, not enough to treat lots of colonies. We therefore decided to undertake further experiments with the ash-D8.

A sufficient number of mites are obviously required in order to be able to carry out the ashing. These are collected with a varroa screen that is inserted under the combs in the brood chamber at the end of June. From then onwards, dead varroa mites drop from the varroa-infested colonies. The mites can be easily stored in a honey jar with a lid. The mites are oval shaped, approximately 1 × 1½ mm in size and red-brown in colour. They dry very quickly and therefore keep for months without decomposing. Between June and July up to ten mites per day fall through the varroa mesh from a serious infestation.

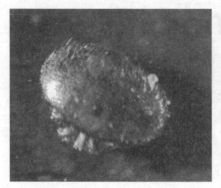 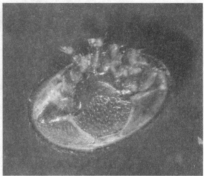

FIGURES 55 AND 56. *Left: Varroa mite dorsal view. Right: Varroa mite ventral view.*

> *The time for doing the ashing is when the sun and moon are in Taurus, around the time of the new moon in May/June.*

For ashing, at least a teaspoonful of mites and completely dry wood are required. Spruce and beech have proven best for this. The mite sample should not contain dead bees or bee parts such as legs or wings, otherwise the bees themselves could be damaged by the treatment. As regards the wood, it is a matter of getting good, firm embers. Some wood species burn very quickly but do not give firm embers. Gas, oil or electric incineration are no use: we have tried them without success. The resulting ash does not have the same effect as ash produced on a wood fire. Furthermore, a stove is needed in which to do the ashing. The stove should be closed on all sides; open fires, such as domestic open fires or bonfires, are unsuitable. An old kitchen stove or old-fashioned room stove are especially suitable. The firebox should be smaller than 25 × 25 cm (10 × 10 in) otherwise the ratio of wood to varroa ash is changed too much. Old cast iron stoves represent the best choice. The flue must be so constructed that it can be closed.

The time for doing the ashing is when the sun and moon are in Taurus, around the time of the new moon in May/June. If an eclipse occurs during this time, the incineration should take place before the eclipse peaks. Various experiments have shown that the life-inhibiting

Figure 57. A wood-burning kitchen stove can also be used for ashing.

Figure 58. Friedrich, our 5-year-old apprentice beekeeper, beside our little stove which has a simple flue support and spark prevention grid. It is very practical, fits in any car boot and is highly suitable for ashing varroa and other pests or weeds.

FIGURES 59 AND 60. Varroa are placed in paper on the embers of a barbecue and then covered with the lid. It is important that the glowing wood embers with the mites are allowed to slowly burn through to form a light-grey ash. The stones are used to restrict the fire in order to control the ratio of wood ash to varroa ash.

effect is strongest before the peak of a solar eclipse. This makes the effect of the ash stronger.

At the appropriate astronomical constellation, we light the wood fire in the stove. Ideally, the ashing should be done at the apiary, but be sure to put a spark guard mesh on the flue to avoid risk of fire. Experiments with other pests have shown that the best effect of the ash was at the location where the incineration took place. The fire and ash boxes should first be thoroughly cleaned out so that there is no longer any ash or dirt in them. There must be enough red embers in the stove to completely cover the grate on which the wood is burning. Now close the flue and place the varroa mites in a paper bag on the embers and close the stove's door. The mites get incinerated. Leave the wood and varroa embers in the stove to turn into ash. The ash should be grey-white, not black, which would only be charcoal and not have the desired effect. After a few hours, remove the cooled down ash from the stove. All the ash from the wood and the mites should be collected together. Put the wood-mite ash in a mortar and grind it for an hour. If you do not have a mortar you may use a

FIGURE 61. *The sieved varroa-wood ash is ground for an hour in a mortar.*

porcelain bowl and a potato masher for grinding. By grinding the ash it gets finer and finer so that it appears to diminish. As the ash gets finer it becomes sticky and adheres to the mortar wall. It can be scraped down with a wooden splint for further grinding. After an hour of grinding, the treatment can begin.

Potentising the ash

To achieve good results, it has become our usual practice to sprinkle the potentised ash between the combs, if possible, at each colony inspection. We are very satisfied with the results. The following explains how to potentise the ash to a D8 potency.

One part of varroa ash is added to nine parts of pure wood ash (the diluent) in a mortar. If you do not have a balance or measuring cylinder, you can use a thimble or an egg cup as an approximate unit of weight or volume. The mixture is ground for three minutes.

In order to keep your mind on the job while potentising, I suggest speaking aloud what you are doing. This avoids, for example, an unexpected visitor or interruption making one forget, for what is said out loud stays in one's ear and is more easily remembered. For example, at the beginning of potentisation we say, 'That will be the D1.' After three minutes grinding, we say loud and clear, 'That is the D1.'

Now we have ten units of D1. Then 90 units of wood ash or dilution material are added to the mortar, followed by the 10 units of D1, and they are ground for three minutes. Again we say, 'That will be the D2,' and at the end of the three minutes, 'That is the D2.' Some readers may think this absurd, but it helps us concentrate properly on the work. This method is continued up to the D8 potency.

At around D4 the container for grinding is no longer big enough. We need to keep in mind that all the ash in the mortar has the same potency, and therefore the same effect. If we again take one part of this and add it to nine parts pure wood ash, we get a D5 potency. Store any ash, whether D3, D4 or D5, in a jar with a lid. The ash can be stored for several years and used for further potentisation. Write the precise potency, date and type on a piece of paper or wood and put it in the jar, for example: 'Varroa D4 June 2020'.

The completed D8 may now be sprinkled from a salt cellar between the combs of the bee colonies. The mites cannot be eradicated with D8. In fact, that would not be our approach.

This treatment creates in the bee colony an atmosphere that is uncomfortable for the mite, and which reduces its reproduction. If, however, there is already a stronger infestation in June, it should be treated with formic, lactic or oxalic acid, for example. Help with these treatments can be obtained from experienced neighbouring beekeepers, beekeeping associations or vets.

The application of ground (dynamised) ash

The most favourable time to apply the ash is with either the moon and sun in Taurus, or with only the moon in Taurus. The application must be done three times. Only then can we count on getting an optimal effect. Up until now we can say that the best results are when the dynamising (the ashing and grinding) and application are done with the sun and moon in Taurus. After four weeks, with the moon again in Taurus, the second application is done, and the third after a further four weeks when the moon is in Taurus again.

The question always raised is why three applications are necessary for the various preparations to be effective. It is not easy to give an exact answer to this. In the experiments done with potencies, ashes and also with the biodynamic spray preparations, it was shown that the results obtained by one or two applications were not as good as those obtained with three applications. In the case of the biodynamic preparations and the potencies and ashes, since we are moving away from the purely material (you need only imagine how little ash material is given to the colonies), we can probably say that we are on a plane where other forces are at work.

The ash is poured into a salt cellar. Ours is made of glass, about 6 cm (2½ in) high and 4 cm (1½ in) wide and has a metal cap with holes drilled in it. The holes of pepper pots are too small. The colony is opened, smoked a little so the bees know that the beekeeper has arrived, and a very small, hardly visible, amount of ash is sprinkled between the frames onto the brood nest. So as not to sprinkle too much ash, it is best to practise beforehand over a sheet of white paper. Close up the hive again and move to the next colony. The quantity of ash that fits in our salt cellar suffices for complete treatment of 25 colonies.

I would just like to give an example that shows how difficult it can be to cure a disease by using the influences of forces

FIGURE 62. *The ash is sprinkled over the brood nest with a salt cellar, so that as much of it as possible falls between the combs.*

whose effects are hardly understood. A beekeeper friend with varroa mites in his apiary reported that with one colony the varroa-D8 preparation (this is the former application method) had shown almost no effect. This left him no peace until he thought of investigating the siting of his apiary by dowsing. He knew that his predecessor had set up the apiary according to particular lines that he found by dowsing. To his great surprise he found that the colony, which hardly responded to D8, was on a crossing point of two lines. This colony was indeed always strong and produced good harvests, but despite that it had a varroa infestation which could not be held in check by D8. From this we can see that certain lines on the one hand can show themselves as very positive, whereas on the other hand they can apparently weaken a colony or even help varroa thrive. The effect of the ground ash was insufficient here. A colony should not be sited in such a position. Such subtleties may help humans in caring for the bee or sap their courage and determination for bee-friendly beekeeping.

Varroa and drones

I have already mentioned that varroa prefers drone brood. This led to beekeepers regularly removing drone brood from colonies as soon as it was capped. However, the desired effect that this would reduce mite numbers largely failed to appear. This might be attributed to the fact that in the early part of the year the ratio of worker to drone brood is unfavourable.

To make best use of the mite-attraction effect of drone brood, a drone comb should be inserted as early as possible in the winter cluster so that drone brood can be created in the colony. This drone comb should be inserted directly into the brood nest in the autumn before the winter feeding, so that in the early spring it is already available for laying. However, at the first spring inspection of the colonies we should pay special attention to the drone comb, because the first drone brood to be laid will have attracted lots of mites. The drone brood should be cut out before it hatches so as to rid the colony of the mites. This method renders a large fraction of mites overwintering in the colony harmless.

The drone comb given in the autumn should previously have had brood in it, because white comb that has never been laid in is less quickly laid in the spring than comb that has already had brood in it. The drone comb is inserted before feeding at the position in the brood nest where the colony is likely to be in the spring after it has consumed its stores. This is generally distant from the entrance, behind the brood nest in a trough hive, or in the upper chamber above the brood nest in a magazine hive.

FIGURE 63. Drone with deformed wings due to varroa infestation at the pupal stage.

To explain why the varroa mite prefers drone brood, we turn to Rudolf Steiner's comments on the individual casts in the bee colony. He described the queen and workers as 'sun animals' and drones as 'earth animals' (*Bees*, p. 9). Looking at the cell shapes that the individual casts develop in, we see that the developing queen has a round cell shape. In contrast, workers and drones have hexagonal cells. As already mentioned, Rudolf Steiner indicated that the hexagonal cell has a connection to quartz crystal with regards to the activities of its forces. Whilst it is true that quartz crystals have very marked 'light forces', they are formed in the ground. Through this formation in the ground the crystal takes up influences that are only in the earthly realm. It can be said that these earthly influences express themselves in the shape of the quartz crystal (*Bees*, p. 48ff).

The relationship between the earthly forces of the crystal and the cells of workers and drones lies not only in their shape, but also in the constellations that favour comb building and cell construction in bee colonies. As has been mentioned already, these constellations govern earth/root times. This tells us that hexagonal cell construction should be regarded as being connected with the activities of earthly forces. From this we can say that the worker bee too is an 'earth animal'. However, as the worker bee requires a different developmental period from that of the drone, it is assigned to the sun, and the crystal shape probably exerts on her the light forces that crystals possess. This illustrates the capacity of the hexagonal shape to exert its influences. Artificial queen breeding gets its material from hexagonal cells. The forces active in the hexagonal cell have exerted their influence on the larva for four or five days, meaning that earth forces have also been absorbed by the larva. The queens that result from these larvae are therefore no longer purely creatures of the sun but have absorbed certain earthly influences.

In view of the many generations of queens that have already been bred from worker larvae and thus have repeatedly taken up earthly forces, it is easy to imagine that their progeny are also stamped with these influences. This means that workers that are produced through constantly repeated breeding over decades, have taken up from the female side, from the queen, an excess of earthly forces, but too few sun forces.

The result of this is that, as regards the actions of forces, worker bees have become more like the earth-animal drones. This perhaps gives us the answer to the question of why not only drone brood but also worker brood is infested by varroa mites. Worker brood has taken up so much of the earthly forces that the mite, which initially infested the earth-animal drone, can now develop in it.

Chapter 13
The Cultivation of Plants for Bees

Landscape can be divided into two types: natural landscape and cultural landscape. In natural landscapes plant growth continues all year, at least when frost is absent, and the availability of flowers continually increases and decreases. In a wonderful way, nature provides each of the various insect groups with flowers as their basis for life. However, for the insect world, these paradisal conditions prevail only where the natural world is not overpopulated by humans. Wherever people settle in large groups and start to live off what nature provides, the time inevitably comes when nature's gifts no longer suffice, and the natural landscape has to be cultivated, thus turning it into a cultural landscape.

Cultural landscape is characterised by the fact that individual plants and their fruits may be abundant, whereas others are entirely lacking. This indicates that while humans can grow specific food plants, we cannot compete with the wisdom of nature or creation. We are often unable to maintain a harmonious balance, but instead tend to go to extremes that work against the balance present in nature.

The result of an artificially created landscape is, among other things, an impoverishment of floral diversity. But because bees cannot live without flowers, beekeepers are forced either to practise migratory beekeeping from spring to autumn, or, in flowerless periods, give their bees appropriate protein in the form of transitional feed. All sorts of pollen substitutes are used for this purpose, but none of them reach the quality of natural pollen. Certainly, the easiest solution for the beekeeper is to feed bees with the necessary proteins. However, beekeepers probably do not have much awareness of the extent to which pollen substitutes affect bee health. From a material point of view a substitute might of course resemble pollen, but it remains a substitute nevertheless.

The result of an artificially created landscape is, among other things, an impoverishment of floral diversity.

In order to provide the bees with natural pollen throughout the year, cultivation of specific flowering plants is indispensable. Such flowering plants include not only trees and shrubs but also the very plants that we can grow in our gardens or fields. The advantage of trees and shrubs is that after planting they do not need so much care. Furthermore, it is easier to find space for them because arable areas of equal size are unnecessary. In contrast, plants grown in gardens or fields generally need constant care, as well as larger areas to make the cultivation worthwhile for the bees. More recently, bees and beekeepers received help through various agricultural support schemes, and now rape, flax, buckwheat and phacelia are increasingly common in regions where they were unknown in the past.

FIGURES 64 AND 65.
Top: Sunflower.
Bottom: Hollyhock.

As most beekeepers have only a garden and not whole fields, it is always somewhat difficult for them to spread bee forage over a large area. It might be possible in arable regions to arrange with individual farmers for them to grow bee plants as an intercrop, thereby benefiting both soil and bees. In regions with lots of fallow land, several beekeepers could club together and develop the land into high quality bee forage areas. But we should not underestimate our own gardens. These days every single flowering plant is of great value to the bees. Native plants above all are important, as was discussed in the previous chapter.

If we have the opportunity to grow bee plants of whatever kind, it is advisable first of all to choose plants already present in the surrounding landscape. If, for example, we want to plant maple we should use the native species and not for decorative reasons some ornamental maple, as this will not provide the expected support for the bees. Furthermore, we should enquire about the soil composition so as to grow plants that really flourish in it. Even the situation of the areas to be cultivated should be considered. On dry slopes only plants that can tolerate drought will grow. In wet valleys, plants should be used that prefer moist conditions.

Once these individual factors have been taken into account, the possibility is created, through making use of cosmic influences, of favourably affecting the pollen and nectar production of annual and perennial plants grown in fields or gardens. Experiments over several years have shown that the pollen yield is increased by sowing, tending and harvesting at light/flower times. When we wish to favourably influence nectar production, warmth/fruit times are chosen. If for various reasons we have not been able to keep to the aforementioned effect at sowing time, we can reactivate it by using the biodynamic spray preparation 501 – horn silica – at light/flower times or warmth/fruit times.

As the following experiment showed, the silica preparation clearly influences the plant's release of nectar and pollen. Half a hectare (1¼ acre) of mustard was sown at light/flower times. Plots were sprayed with the silica preparation at each of light, water, warmth and earth times and corresponding control plots were left untreated. The plot treated at light/flower times was the one most frequently visited for pollen, whereas the plot treated on warmth/fruit times was visited predominantly for nectar.

If we grow plants for providing pollen and nectar, it is advisable to spray part of them with horn silica preparation at light/flower times, and a part at warmth/fruit times. But if we want to increase the nectar or the pollen availability, we can spray the whole area at light/flower or at warmth/fruit times. If we intend the plants we are growing to provide seed for the following year, the harvest should likewise take place at light/flower or at warmth/fruit times, because the time of harvest very strongly influences the properties of the plants in the following year. As different plant species feel at home in the different soil types and different climatic regions, it is not possible at this point to offer a list of forage plants. In the individual localities, it is advisable to enquire at the flower and tree nurseries and with local farmers to see which plants flourish best and can therefore be grown successfully.

The aim in growing bee plants should be to guarantee, through increased flower numbers, a supply of pollen and nectar through spring, summer and autumn, in order to make any interim feeding of sugar and/or pollen substitutes unnecessary. The scope of bee plant cultivation does not need to be so great as to contribute to a harvest, but should enable the bees to keep gathering pollen and nectar and to continue raising brood.

Figures 66–69.
Top left: Phacelia.
Top right: Rosebay willowherb (fireweed).
Bottom left: Borage.
Bottom right: Buckwheat.

Chapter 14
The Conservation of Bees for the Future

In view of both the current environmental situation and the difficulty in obtaining healthy food, we might consider more closely the significant role that bees play in the natural order of things and how we can conserve them for the future.

We know that plant diversity decreases when bees are absent for a long time. Not only is this brought about by the lack of pollination, but also as a result of the lack of the poison-like substances that bees secrete before taking nectar and which the plants find so beneficial. The provision of these substances and pollination are necessary for the survival of plants, and we cannot imagine the earth without the plant world.

Then there is the honey that bees produce and that we enjoy. Some people might think that the production of honey is less important than the bees' role as pollinators. But does anyone know what it would mean for humanity if bees no longer produced honey?

Individual fruits or substances leave behind a corresponding effect in humans. The human organism can manage without the effect of these individual substances for a while because it can create a balance over time. But if, for example, flower honey was withdrawn for a long time, and with it the hexagonal crystalline forces it contains, then bone and musculature growth may suffer as a result of the weakening blood. According to Rudolf Steiner, honey helps to shape and maintain the form of the human body, and as a mother's milk is to the suckling child, so too is honey to the elderly. Humans can live to a limited extent without these two substances, but we probably cannot do without them completely.

According to Rudolf Steiner, honey helps to shape and maintain the form of the human body, and as a mother's milk is to the suckling child, so too is honey to the elderly.

It is clear from this how difficult it is to rank the importance of an organism's individual functions, as all its functions, even apparently minor ones, fulfil their purpose and keep it alive. The bee is part of the living organism of the earth, and this organism also needs human beings in order to maintain itself.

With this in mind, it becomes clear what tasks fall to beekeepers. At present bees are regarded just as producers of natural products. Through various procedures they are pushed to ever higher productivity, which leaves them considerably weakened and susceptible to disease. This purely economic exploitation of bees has resulted in our placing them on the same level as our unbalanced factory farming. It is known that factory farming, through its sophisticated breeding and by restricting animals to the least possible space, has led to the increasing spread of disease. We can now see a similar problem with the bees, which has become especially clear with the rapid spread of the varroa mite.

As a result of bees drifting to apiaries a long way off (rather than just within apiaries as was previously thought), the varroa mite was able to spread over large areas. Other diseases, such as foulbrood, spread among colonies within an apiary also as a result of drifting. The question thus arises as to what extent large apiaries are sustainable in the future. To eliminate the rapid spread of disease, or at least to restrict it as much as possible, we would have to return to setting up small apiaries, which we would then scatter in the landscape in such a way that drifting could largely be avoided.

As part of the big task that humankind faces regarding the preservation of bees, bee drifting of course has a subordinate role. The focus will have to be directed somewhere else. In this book it has been constantly indicated that we must recognise and nurture the essential nature of bees. It will occur to individual beekeepers to devote themselves wholly to their bees and the bees' way of life, and put aside their own views and wishes regarding beekeeping for profit. If they succeed in keeping their bees in such a way that the care they give is beneficial and bee-friendly, the issue of profitability will solve itself. Of course, one or two beekeepers might hold the view that much of what is presented in this book represents a step backwards for beekeeping. But if what I have written here succeeds in helping beekeepers return their bees to a healthy robustness, it will be a step into the future.

This thoroughly unconventional way of accompanying bees into the future will be much easier for those to whom bees and honey have once again become something special.

Rudolf Steiner's indications regarding bees, and the involvement of cosmic influences in practical beekeeping, will greatly facilitate this path and bring us closer to the goal of bee-friendly beekeeping. However, this goal is attainable only if as many beekeepers as possible adopt this course of action.

Bibliography

Steiner, Rudolf (1993) *Agriculture*, Biodynamic Association, USA
Steiner, Rudolf (1998) *Bees*, Anthroposophic Press, USA.

Recommended Reading

Bees and Beekeeping

Bresette-Mills, Jack (2016) *Sensitive Beekeeping: Practicing Vulnerability and Nonviolence with your Backyard Beekeeping*, Lindisfarne Books, USA.
Hauk, Gunther (2017) *Toward Saving the Honeybee*, Biodynamic Association, UK.
Kornberger, Horst (2019) *Global Hive: What the Bee Crisis Teaches Us About Building a Sustainable World*, Floris Books, UK.
Weiler, Michael (2019) *The Secrets of Bees: An Insider's Guide to the Life of Honeybees*, Floris Book, UK.

Biodynamics

Klett, Manfred (2005) *Principles of Biodynamic Spray and Compost Preparations*, Floris Books, UK.
Koepf, Herbert H. (2012) *Koepf's Practical Biodynamics: Soil, Compost, Sprays and Food*, Floris Books, UK.
The Maria Thun Biodynamic Calendar, Floris Books, UK.
Masson, Pierre (2014) *A Biodynamic Manual: Practical Instructions for Farmers and Gardeners*, Floris Books, UK.
Pfeiffer, Ehrenfried E. (2011) *Pfeiffer's Introduction to Biodynamics*, Floris Books, UK.
Pfeiffer, Ehrenfried E. (2012) *Weeds and What They Tell Us*, Floris Books, UK.

Index

anti-bird nets, removing 15
American foulbrood
   (malignant foulbrood)
   152
aphids 126, 128, 139
artificial swarms 56
ashing of pests 164,
   169–71

beekeepers, neighbouring
   53
benign foulbrood
   (European foulbrood)
   156
Berlepsch, August von 105
breeding colony 92, 95
brood combs 31–35
building frame 37–39, 41

cell
— construct 39f
—, hexagonal 61
—, round 61
cell-moulding tool 93
chalkbrood 49, 156f
chamomile 147f
clamp plugs
   (*Klemmstopfen*) 74
clearing flight 15–17
cold-way (comb hanging)
   81
comb
—, age of 118f
— removal 132

dandelion 148
drone
— and varroa 33
— brood 176, 178f
— cells 104, 123
— comb 41
dysentery 158

earth/root times 25f
European foulbrood
   (benign foulbrood) 156
extractor room 133
eyes 48

feed 66, 77
field bee 125
flower honey 138f
flying bees 125f
fondant 33
— plug 82
forest honey 127
foulbrood 152–54, 156
foundation comb 111, 116
— wax 115

Gerstung, Ferdinand 109
grafting tool 74, 95

hatching calculation 98
hexagonal cells 103
hive
— layout 121
—, preparing 65

honey 133
—, crystallised floral 143
—, different kinds of 140
— flavour 134
—, granulated 134, 136
— chamber 121f
— harvest 28
— processing 131–33,
   135–37
— storage (in hive)
   122–24
—— (after extraction) 137
honeydew 127
— honey 138–40, 143
horizontal hives (trough
   hives) 32
horsetail 148f
house bee 125

inspection of colonies 26
internal mite (tracheal
   mite) 159

Kirchhain Mating Nuc
   Box (KMNB) 81
Klemmstopfen (clamp
   plugs) 74

larvae grafting 27, 95f
light/flower times 25

magazine hives 32, 37, 122
malignant foulbrood
   (American foulbrood)
   152

Marburg box 54–57
mating
— colony 83
— flights 89
Mehring, Jean 105, 111
melicitose honey 130
molasses 143, 158
mouse guards, removing 15

Nasonov scent glands 64, 69
natural comb 112f, 116
— construction 105
nectar 126–28
—, dew 126–28
—, flower 126
nettle 148
nosema 160f
nuc 17, 49, 70–73
— with capped brood
   80, 82f
— box 81

oak bark 148f
ocelli 48
orientation landmarks 72

Perkiewicz, Ernst 50, 54,
   153
pest, ashing of 164,
   169–71
phloem 128
pollen storage 122–24
potentisation 164, 173f
propolis 65

quartz crystal 131
queen
—, breeder 61
— breeding 74, 88f,
   92–94, 96, 98
— cage 85f
— cells 94, 98f, 104, 108,
   122
— cups 41–43, 46, 93, 96
—, emergency 60
—, emerging of 67
— excluder 35, 122
—, rejection of 78
—, supersedure 59
—, swarm 59
—, tooting of 67f, 98
queen-right colonies 99

rear-access hives 20, 37
rearing colony 92–94

skep (swarm box) 63,
   143–45
spring inspection 31
spring sickness 159
starter-strips 113
Steiner, Rudolf 7
stonebrood 157f
strainer 133
swarm
—, artificial 75–80
— box (skep) 63, 76
— de-queened 69
— handling 62–64

swarming 43–55, 61
swarm-resistant bees 45, 57

thoroughbred bees 89f
Thun, Maria 8
tooting of queen 67f, 98
top-access hives 37, 122
tracheal mite (internal
   mite) 159
trough hives (horizontal
   hives) 32, 122

uncapping tool 133

valerian 148
varroa mites 162f, 167,
   169f, 176, 178f
— and drone brood 33
varroosis 162–65

warmth/fruit times 25
warm-way (comb
   hanging) 81
water/leaf times 25
water supply 71
wax 104
winter feed 142–44, 146f,
   158
winter inspection sheet 15f
worker bee 125
— cells 104, 123f

yarrow 148

Floris Books

For news on all our **latest books**,
and to receive **exclusive discounts**,
join our mailing list at:

florisbooks.co.uk

Plus subscribers get a FREE book
with every online order!

We will never pass your details to anyone else.